Extinction and Memorial Culture

This book considers how we encounter and make meaning from extinction in diverse settings and cultures. It brings together an international and interdisciplinary range of scholars to consider how extinction is memorialised in museums and cultural institutions, through monuments, in literature and art, through public acts of ritual and protest, and in everyday practices.

In an era in which species are becoming extinct at an unprecedented rate, we must find new ways to engage critically, creatively, and courageously with species loss. *Extinction and Memorial Culture: Reckoning with Species Loss in the Anthropocene* develops the conceptual tools to think in complex ways about extinctions and their aftermath, along with providing new insights into commemorating and mourning more-than-human lives.

This book will be of great interest to students and scholars of the environmental humanities, extinction studies, memorial culture, and the Anthropocene.

Hannah Stark is Associate Professor in English at the University of Tasmania, Australia. She is the author of *Feminist Theory After Deleuze* (2016), the co-author (with Timothy Laurie) of *The Theory of Love: Ideals, Limits, Futures* (2021), and the co-editor of *Deleuze and the Non/Human* (2015), and *Deleuze and Guattari in the Anthropocene* (2016).

Extinction and Memorial Culture

Reckoning with Species Loss in the Anthropocene

Edited by
Hannah Stark

Routledge
Taylor & Francis Group
LONDON AND NEW YORK

First published 2023
by Routledge
4 Park Square, Milton Park, Abingdon, Oxon OX14 4RN

and by Routledge
605 Third Avenue, New York, NY 10158

Routledge is an imprint of the Taylor & Francis Group, an informa business

British Library Cataloguing-in-Publication Data
A catalogue record for this book is available from the British Library

Library of Congress Cataloging-in-Publication Data
Names: Stark, Hannah, 1983– editor.
Title: Extinction and memorial culture : reckoning with species loss in the Anthropocene / edited by Hannah Stark.
Description: Abingdon, Oxon ; New York, NY : Routledge, 2023. | Includes bibliographical references and index. | Contents: Introduction : reckoning with species loss in the Anthropocene / Hannah Stark—Layering loss : a conversation with Lucienne Rickard / Keely Jobe—Beginning and endling : the case of Aotearoa New Zealand's first Rose-Crowned Fruit—Dove, or Specimen OR. 030538 / Rosie Ibbotson—Listening to lost species : memorialising extinction through sound / Hannah Hunter—Entangled extinctions and cultural resurgence / Susanne Ferwerda—Edenic extinction : memorialising lost species across timescales at the Eden Portland Project / Clara de Massol—Franklinia in the garden : memorialising foliage, preserving heritage / Kelly Enright—Withnessing : multispecies approaches to extinction, testimony, and bodies of water / Toby Juliff—Conjuring up ghost species : on photography and extinction / Shane McCorristine, William M. Adams and Adam Searle—An elegy for an ecotype : Eva Saulitis's Into Great Silence and the extinction of the Chugach Transient Killer Whales / Jennifer Schell—Lost species, lost worlds : memorialising extinction in the art of Todd McGrain and Chris Jordan / Hannah Stark—Psittacine extinction story : I once loved a lorikeet / Hélenè Frichot. |
Identifiers: LCCN 2022061408 | ISBN 9781032326375 (hardback) | ISBN 9781032326382 (paperback) | ISBN 9781003315957 (ebook)
Subjects: LCSH: Extinction (Biology)—Social aspects. | Extinct animals in art | Memorialization. | Human ecology.
Classification: LCC QH78 .E94 2023 | DDC 576.8/4—dc23/eng/20230417
LC record available at https://lccn.loc.gov/2022061408

ISBN: 9781032326375 (hbk)
ISBN: 9781032326382 (pbk)
ISBN: 9781003315957 (ebk)

DOI: 10.4324/9781003315957

Typeset in Times New Roman
by codeMantra

Contents

Contributors

William M. Adams retired from the Moran Chair of Conservation and Development at the University of Cambridge in 2020. He is a geographer, and currently teaches at the Geneva Graduate Institute. His work focuses on conservation and society from the perspectives of political ecology and environmental history, and he has a particular interest in the significance of digital technologies for ideas about nature. His book, *Strange Natures: Conservation in the Era of Genome Editing* (with Kent Redford), was published by Yale University Press in 2021.

Clara de Massol is a lecturer at King's College London, where she completed her PhD on memory in the Anthropocene. Her research, informed by her background in cultural and literary studies, examines the intersections of memory studies and environmental humanities. Her writing more particularly investigates the ways in which memorial forms and new modes of remembrance emerge and are formulated in the Anthropocene. Clara's doctoral research was supported by the London Arts and Humanities Partnership (AHRC), and she was the 2019 recipient of the Memory Studies Association's Excellent Paper award.

Kelly Enright holds a doctorate in History from Rutgers University and a master's in Museum Anthropology from Columbia University. Her research focuses on human-animal relationships, cultural constructions of nature, and museum history. She is a former fellow at the Animals & Society Institute and founding co-editor of *Sloth*, a journal for early-career scholars in Human-Animal Studies. Enright is author of *Rhinoceros* (Reaktion), *The Maximum of Wilderness: The Jungle in the American Imagination* (University of Virginia), and *Osa & Martin: For the Love of Nature* (Pequot). She has also published articles for both academic and general audiences, including work in *Environmental History, Cultural Studies Review, National Geographic*, and *American Archaeology*. She is Associate Professor of History and Anthropology, and Director of Public History, at Flagler College in St. Augustine, Florida, where she is actively engaged in public and digital history.

Susanne Ferwerda is a lecturer in Comparative Literature at Utrecht University, the Netherlands. She holds a PhD in English from the University of Tasmania, Australia. *Blue Ocean Stories*, her first book project, analyses anticolonial and decolonial narrative disruption to Western colonisation in the Oceanic region via long and short fiction, and visual and performance art.

Hélène Frichot is Professor of Architecture and Philosophy, and Director of the Bachelor of Design, Faculty of Architecture, Building and Planning, University of Melbourne, Australia. In her former position, she directed Critical Studies in Architecture, at the School of Architecture, KTH Stockholm, Sweden. Her recent publications include *Dirty Theory: Troubling Architecture* (AADR, 2019), *Creative Ecologies: Theorizing the Practice of Architecture* (Bloomsbury, 2018), and *How to Make Yourself a Feminist Design Power Tool* (2016). She is editor of a number of collections, including with Catharina Gabrielsson and Helen Runting, *Architecture and Feminisms: Ecologies, Economies, Technologies* (Routledge, 2017), with Naomi Stead, *Writing Architectures: Ficto-Critical Approaches* (Bloomsbury, 2020), with Marco Jobst, *Architectural Affects After Deleuze and* Guattari (Routledge, 2021), and most recently with Adrià Carbonell, Hannes Frykholm, and Sepideh Karami, *Infrastructural Love: Caring for Our Architectural Support Systems* (Birkhauser, 2022).

Hannah Hunter is a PhD candidate at the Sonic Arts of Place Laboratory in the Department of Geography and Planning at Queen's University, Canada. Her research interests include sound studies, animal geography, science and technology studies, and environmental history. Her doctoral work explores themes of birds, sound, and extinction in North America, with a focus on how bioacoustics, field recording, and sound art have all responded to, and been affected by, the avian extinction crisis. Hannah's PhD is funded by a Vanier Canada Graduate Scholarship, and additional fieldwork support has been provided by a Mitacs Globalink Research Award.

Rosie Ibbotson is Senior Lecturer in Art History and Theory at Te Whare Wānanga o Waitaha University of Canterbury, Aotearoa New Zealand. Ibbotson's research concerns the entanglements of images, museums, and environmental violence, and she is currently writing a book titled *Picturing the imperial Anthropocene: Visual Representation and Environmental Change in Long Nineteenth-Century Aotearoa New Zealand.* Ibbotson also has research interests in the field of animal studies, and she has published on the intersections of museology, visual and material culture, and de-extinction.

Keely Jobe is a doctoral candidate in English at the University of Tasmania. Her thesis focuses on lesbian separatism and feminist evolutions, and her broader research interests include creative writing, feminist philosophies, queer ecologies, extinction studies, and plant studies. Keely's work has been published in *The Monthly, Island Magazine, The Living Archive: Extinction Stories from Oceania*, and the *Tasmanian Land Conservancy's* "Breathing Space".

Toby Juliff is Lecturer in Art at the School of Creative Arts and Media, University of Tasmania. Previously lecturer at the University of Melbourne and Leeds Arts University, Toby has published essays and chapters on subjects as diverse as exhibition histories of British art in the 1960s, participatory art and heritage studies, confessional video art, and most recently, Latinx feminist sculpture. His most recent project – entitled Posthumanism at the Edge of the World – explores new "Southern" geographies and terrain for post-Anthropocene thinking and making in contemporary art.

Shane McCorristine is an interdisciplinary historian at Newcastle University in the UK with interests in the "night side" of modern experience – namely social attitudes towards death, crime, dreams, ghosts, and the supernatural. He is the author of *The Spectral Arctic: A History of Dreams and Ghosts in Polar Exploration* (UCL Press, 2018), and *Spectres of the Self: Thinking about Ghosts and Ghost Seeing in England, 1750–1920* (Cambridge University Press, 2010), and the editor of *Interdisciplinary Perspectives on Mortality and its Timings; When is Death?* (Palgrave Macmillan, 2017). He is currently involved in a research and public engagement project on the history of Newcastle Gaol (1828–1925) – its inmates, executions, and role in the evolution of crime and punishment in Newcastle-upon-Tyne.

Jennifer Schell is Professor of English at the University of Alaska Fairbanks. Her specialties include North American literature, circumpolar writing, extinction studies, and environmental humanities. She is the author of *"A Bold and Hardy Race of Men": The Lives and Literature of American Whalemen* (University Massachusetts Press, 2013). Since then, she has been working on a series of articles that address representations of anthropogenic species extinction in North American print culture. At present, she is drafting a book manuscript entitled *Ghost Species: North American Extinction Writing and the Ecogothic.*

Adam Searle is a Research Fellow in the School of Geography, University of Nottingham, and his research examines the relations between humans, other animals, and technologies. He was initially an ecologist before completing a PhD in Geography at the University of Cambridge in 2021, and was a postdoctoral fellow in science and technology studies at the Université de Liège in Belgium. In 2021, he co-founded the Digital Ecologies research group, and he is a co-editor of *Digital Ecologies: Mediating More-than-human Worlds* (Manchester University Press, 2023).

Hannah Stark is Associate Professor in English at the University of Tasmania. She uses theories at the forefront of the humanities to advance our understanding of contemporary social and political issues such as extinction and climate change, the emergence of public environmental sentiment, marriage equality, sexual politics, and post-nuclear forms of kinship and care. Hannah is the author of *Feminist Theory After Deleuze* (Bloomsbury, 2016), the co-author (with Timothy Laurie) of *The Theory of Love: Ideals, Limits, Futures* (Palgrave Macmillan, 2021), and the co-editor of *Deleuze and the Non/Human* (Palgrave, 2015), and *Deleuze and Guattari in the Anthropocene* (Edinburgh, 2016). She co-edited special issues of *Australian Humanities Review*, *Parrhesia: A Journal of Critical Philosophy*, and *Deleuze Studies.* Hannah is currently lead-investigator on an Australian Research Council funded project "Beyond Extinction: Reconstructing the Thylacine (Tasmanian Tiger) Archive".

Acknowledgements

I acknowledge and pay my respects to the Muwinina people, the original custodians of the land on which I work, who did not survive colonisation, and to the Palawa people, who care for Country today. I pay my respects to elders past, present, and emerging and to those who never made elder status. Considering colonial legacies is particularly significant for a book about extinction. In lutruwita/Tasmania, where I live and work, the thylacine became extinct just 133 years after European invasion in a shocking indictment of the capacity of European colonists and their descendants to care for and live with endemic fauna. Australia now has the worst mammalian extinction rate in the world. Living in this place has shaped how I experience and think about extinction.

I would like to thank Keely Jobe for her research assistance with this book. Keely has been not only a deeply committed and engaged research assistant but also an important interlocutor for thinking about extinctions of all kinds. I acknowledge the early input of Katrina Schlunke on this project and for my broader considerations of extinction. Thanks to those students who have thought with me about extinction in the Anthropocene, even when it was hard work. Thank you to all the contributing authors. It truly has been a pleasure to engage with your work.

A huge debt of gratitude to Anne for her support, interest, and enthusiasm for this project and for everything else.

This book was supported by Australian Research Council funding.

Reckoning with Species Loss in the Anthropocene

Hannah Stark

In January 2021, in the central hall of the Tasmanian Museum and Art Gallery, Lucienne Rickard erased a large-scale led pencil drawing of a Swift Parrot. She had completed this drawing as part of her multiyear durational performance *Extinction Studies* in which she meticulously drew and then erased species from the International Union for Conservation of Nature's (IUCN) "Red List of Threatened Species". The Swift Parrot had taken five weeks to complete and was erased in a matter of minutes. "Swift parrot, not extinct, approximately 300 left", the artist noted to the gathered public before erasing the image. The atmosphere in the silent room was electric; people cried. In the Bristol Museum in 2019, thirty-two of the extinct and endangered animal mounts in their World Wildlife Gallery were shrouded in black cloths in an exhibition called "Extinction Voices". The erasure or disappearance of these species was represented by the translucent black fabric which also invited museum viewers to see these extinct and critically endangered animals re-presented to the public in black mourning veils. In the United States, Todd McGrain's large bronze sculptures of extinct birds have been placed in the location in which they were last seen in the wild. These site-specific public environmental artworks are sombre public memorials to once common bird species that have now been lost: the Labrador Duck, Heath Hen, Carolina Parakeet, Great Auk, and Passenger Pigeon. Outside the Muséum National d'Histoire Naturelle in Paris in the Jardin des Plantes sits the Dodo Manège, a carousel composed of likenesses of extinct and endangered animals. In contrast to this carnivalesque display, the museum contains the Room of Endangered and Extinct Species which houses hundreds of taxidermy mounts of animals on the edge of extinction or who are already gone. In an Extinction Rebellion march in London in October 2019, protesters in the "skeleton crew" wore black mourning clothes and carried skeletal bone-sculptures. The march was framed as "a ceremony of grief, followed by a funeral procession through the streets of London" and as an opportunity to collectively express "profound grief" for species loss.[1]

These are all examples of how extinction is increasingly being memorialised through art, museum displays, consciousness-raising events, monuments, protests, and ceremonies. They attest to the extinction cultures which

DOI: 10.4324/9781003315957-1

are emerging as part of public discourse and show an increasing public concern for our current extinction crisis. Mass extinction and the diminishment of biodiversity is one of the most significant issues facing our time – a period now widely described as the Anthropocene. This is a geological period in which human impact, particularly the ongoing impacts of colonialism, capitalism, and extractive economies, is shaping the earth in lasting ways. Dipesh Chakrabarty has noted that the Anthropocene signals a collapse of human and natural histories and that the human is now a geological force (15). The 2019 Intergovernmental Science-Policy Platform on Biodiversity and Ecosystem Services' (IPBES) *Global Assessment Report on Biodiversity and Ecosystem Services* is clear that "Human actions threaten more species with global extinction now than ever before" and that "around 1 million species already face extinction, many within decades, unless action is taken to reduce the intensity of drivers of biodiversity loss" (xv–xvi). They also note that the rate of extinction is likely to accelerate. The IUCN has assessed that over 41,000 species are currently threatened with extinction and appear on their "Red List". In 2020 alone, the IUCN declared thirty-one more species extinct. We are currently experiencing what is being called the sixth extinction (see Kolbert). As Richard Grusin writes,

> [u]nlike these others, the sixth extinction is attributed to the agency of humans, or more accurately to the agency of nonhuman processes set in motion by humans, particularly the technologies of industrial capitalism and globalization that continue to operate both in service to and irrespective of human intent and purpose.
>
> (viii)

Memorial culture is shaped by public feelings. Memorials are, Erica Doss writes, "archives of public affect" (13) which tell us something about our cultural values and shared emotions over time. Nowhere is this more evident than in the toppling and removal of public memorial statues. The "Rhodes Must Fall" movement at the University of Cape Town is a pivotal example in which a significant protest movement demanded the removal of the statue of Cecil Rhodes, a former Prime Minister of Cape Colony, imperialist, white supremacist, and someone who restricted the rights of black Africans therefore paving the way for Apartheid. The "toppling" of this statue was a key moment in the decolonisation of university campuses (and other public spaces) in a global context. In my own city of Hobart, Australia, the city council has just agreed to remove the statue of William Crowther after extended and escalating public pressure, particularly from Aboriginal groups. Crowther was the Premier of Tasmania (1878–1879) and is a controversial figure because in 1869 he mutilated the body of William Lanne (falsely believed to be the last full blooded male Tasmanian Aborigine) and sent his head to the Royal College of Surgeons in London. While this was not officially sanctioned – Crowther broke into the morgue and was consequently suspended from his position at the Hobart General Hospital – it did not prevent him from becoming Premier of the island state. These examples show how public memorials shape and manage public narratives about the past while, concomitantly, attitudes to memorial sites and cultures reveal shifting social mores and political sentiments.

This book takes as its objects of analysis memorial culture. It is interested in monuments, and memorials that commemorate extinction, where they can be found, but it is equally interested in the more ephemeral aspects of memorialisation such as those found in cultural institutions such as museums, in photographs, performances, literature, art, performance, and actions. These various spaces in which we reckon with species loss are sites of memory work and where narratives are composed, agreed upon, challenged, and reworked. Eco-mourning has been critiqued for "its putatively backward-looking, nostalgic impulses, as well as for its hegemonic stranglehold over environmental discourses" (O'Key 647). Alternatively, Adrian Parr writes about memorial culture as a form of cultural production which "organizes the energies, affects, and forces of memory" (3). She describes it as "utopian memory thinking: one where culture inhabits the disruptive dimension of traumatic memories which also entails a little bit of forgetting, while simultaneously bringing forth a sense of agency" and that "at its most successful, [memorial culture] avoids monumentalizing the past, choosing instead the tension between past and present, while joyfully looking to the future" (3). Parr is working in a Deleuzian tradition, and within this context, "joy" has a Spinozan resonance and should be understood as the capacity to increase joyful passions and therefore action. In this context, joy is not a synonym for happiness. What Parr is suggesting is that memorial culture need not paralyse us in the present, in grief and hopelessness, but that it has the capacity to galvanise and promote action for a better future. Memorial cultures of extinction create a space and time for mourning species loss, but they are generally didactic and provide us with lessons for the future, a hopeful future in which humans can be less environmentally destructive. The examples considered in this book will illustrate the future-focussed nature of extinction and memorial culture.

As the above examples of Rhodes and Crowther demonstrate, memorial culture usually commemorates human lives and histories. However, the memorial culture forming around extinction can be read as a barometer of public environmental sentiments. It reveals cultural attitudes to animals, plant life, and nature more broadly and shows us how we are coming to terms with human environmental impact in the Anthropocene.[2] Kelly Enright suggests that "[i]n examining how a species lives in memory after its death, cultural conversations about nature and wildness are revealed, and about the placement of animals within stories of human history" (155). There are existing memorials to animals for example, the "Animals in War" memorial, which is a joint project between the Australian War memorial and the Royal Society for the Prevention of Cruelty Against Animals (RSPCA), and which commemorates those animals who served in war and conflict. Another example is the statue of "Trim" on Macquarie Street in Sydney, outside of the State Library. Trim was Matthew Flinders' beloved cat and accompanied him on his circumnavigation of Australia in 1801. However, these animals are commemorated by humans because of their important relationship to our species and events in human history. The significance of memorial cultures of extinction, Dolly Jørgensen writes, is:

> Although monumentalisation is thought of as a practice by humans to remember the loss of other humans, these memory practices stretch to encompass the non-humans in forms from pet cemeteries to warhorse memorials. Because humans are the commissioning agents of these works, studying memorials and the sentiments behind them gives us a glimpse into how those humans think about the relationship between humans (past, present, and future) and animals. Because monumentalisation generally involves honoring the subject, the animal commemorated is honored as worthy of being remembered. At the same time, monuments fix the human-animal relationship in space and time, which may lock the remembering into particular forms that prove relevant for future viewers.
>
> (185)

Memorial cultures of extinction therefore challenge the anthropocentrism of established memorial practices.

Why is it important to memorialise extinction? There is an urgency to our current environmental moment in which species loss is accelerating and so it is critical for us to consider where and how we process the enormity of the extinction crisis as a cultural and political phenomenon that both confounds us and calls us to action. Responding to Deborah Bird Rose's insistence that we "live in a time of almost unfathomable loss, and we are called to respond" (1), Chrulew and De Vos write of the extinction crisis:

> It brings forth new modes of commemoration and mourning, and new practices of archiving and survival. It calls for action in the absence of hope, and for the recognition and nourishment of new generativities: New modes of assemblage and attachment, resurgence and reworlding, commoning, composting and caring for country.
>
> (23)

This book considers how extinction is marked and experienced in a variety of Western contexts. It reflects on what it means to think about extinctions that are caused directly or indirectly by humans. How do we grieve this mess of our own making? What is the reckoning that this grief will bring forth? In engaging with environmental grief and trauma, and cultures of loss and memory centred on nonhuman species, this book endeavours to diversify our commemorative landscape for the future so that we can better acknowledge more-than-human lives and deaths.

This book contains chapters from a range of international contexts and covering diverse approaches to memorialising extinction. The first chapter, "Layering Loss; a Conversation with Lucienne Rickard" is Keely Jobe's interview with the artist about her ongoing durational artwork *Extinction Studies*. Described by Jobe as an "evolving memorial to extinction" (x), *Extinction Studies* involves the artist drawing extinct and endangered animals in meticulous detail only to erase them once they are completed. The interview engages with the context in which Rickard's performance is occurring, the Tasmanian Museum and Art Gallery (TMAG). Through discussing the specificity of this cultural institution, Jobe and Rickard consider the "cultural and ethical reckoning" currently occurring in museums globally. They also consider notions of presence, absence and erasure in memorial culture and cultural memory and the range of emotions that extinction elicits.

In Chapter 2, "Beginning and Endling: The Case of Aotearoa New Zealand's First Rose-Crowned Fruit Dove, or Specimen OR.030538", Rosie Ibbotson writes about her experience of visiting the remains of a specimen, who she calls Ori, at Aotearoa's national museum Te Papa Tongarewa. Working through the unusual history of this creature, the first and last of its kind in New Zealand, Ibbotson examines museological practices in relation to extinction and the preservation, storage, and safeguarding of specimens. More broadly, she considers the complicity of museums with anthropogenic extinction and the "the extent to which archival forms of 'memorialising' loss might seem to normalise rather than mourn it" (x).

In Chapter 3, "Listening to Lost Species: Memorialising Extinction Through Sound", Hannah Hunter considers sonic memorials to extinction. After contextualising the place of sound in extinction memorials with sonic memorials more generally, Hunter offers a close reading of the work of several American and French artists, Steve Norton's *Requiem* (2018), Elizabeth Turk's *Tipping Point: Echoes of Extinction* (2020) Michael Reiley's *Echozoo* (2020), and Marguerite Humeau's *Opera of Prehistoric Creatures* (2012). Examining audio recordings and sound art that engage with extinction, particularly those that "give voice" to extinct animals, Hunter argues for the affective potential of sonic memorialisation. She argues that through cultivating a "novel and powerful connection with lost animals" (x) sonic memorials invite the listener to develop greater empathy. In this way, she considers questions that recur throughout this collection – what is the role of memorialisation in eliciting environmental action? How might mourning extinct animals provoke actions which could prevent future extinctions?

In Chapter 4, "Entangled Extinctions and Cultural Resurgence", Susanne Ferwerda takes us under water and into the kelp forests off the coast of lutruwita/Tasmania. It is here that we encounter creatures such as the endangered Spotted Handfish and Red Handfish, and where the extinct

Smooth Handfish would have once made their home. Ferwerda examines the art of Lucienne Rickard, whose *Extinction Studies* engages with the Smooth Handfish and, as discussed in the opening chapter, who works in and with the complex history of the Tasmanian Museum and Art Gallery and it's complicity with species extinction and the problematic and incorrect narrative about the extinction of the Tasmanian Aborigines. Ferwerda then turns to the work of Indigenous artist Mandy Quadrio and the way in which it addresses the aftermath of colonialism and the ongoing impacts of this on Indigenous culture and marine environments. Focussing powerfully with concepts of erasure, extinction, and resurgence, Ferwarda turns to Quadrio's work with kelp to show the persistence of Indigenous cultural practices today.

In Chapter 5, "Edenic Extinction: Memorialising Lost Species Across Timescales at the Eden Portland Project", Clara de Massol examines Eden Portland, a planned extinction observatory and memorial to be built under the limestone cliffs of Portland, in Dorset in the United Kingdom. Taking a critical approach to this project, Clara de Massol examines the proposed monument to extinction and its ideological approach to the narrative of the extinction crisis. Reading the absence of engagement with the violent extractive history of the proposed site, Clara de Massol examines the ways that Eden Portland decontextualises extinction. While it celebrates biodiversity, Eden Portland's interpretation of extinction narratives remains individualist, neoliberal, and a celebration of human ingenuity. Clara de Massol considers the way this project de-politicises extinction and remains anthropocentric in orientation.

In Chapter 6, "Franklinia in the Garden: Memorialising Foliage, Preserving Heritage", Kelly Enright examines a species which is propagated in public and private gardens, but which has been considered extinct in the wild since its last sighting in 1803. Enright's chapter reveals how the absence of Franklinia is marked in what was its native habitat, the Altamaha River area in the state of Georgia in the southern United States, as well as the place of cultivars of this species in historically significant sites such as John Bartram's estate in Philadelphia. Trees are commonly used in memorial contexts, and Enright is interested in the role of national heritage in horticultural heritage and conservation. Considering themes of memorialisation and survival, Enright ultimately reads the Franklinia growing in gardens as a memorial to itself.

In Chapter 7, "Withnessing: Multispecies Approaches to Extinction, Testimony, and Bodies of Water", Toby Juliff examines the work of artist Selena de Carvahlo. Beginning with de Carvahlo's artist residency in which she (and others) swam with Giant Australian Cuttlefish off the coast of Whyalla in South Australia, Juliff mobilises Dolly Jørgenson's notion of an "energy landscape" to consider the extractive economies of this location. In the engagement between the artist and the Giant Australian Cuttlefish, Juliff finds resources to examine witnessing, testimony and shared witnessing (or withnessing) between humans as a "radical reorientation of human-centred testimony of extinction" (x). Juliff utilises Astrida Neimanis's poetics of "bodies of water", to argue for a hydrofeminist ethics of relation to more than human lives and to connect to his second case study of de Carvahlo's work, *Beware of Imposters (the secret life of flowers).* Connected to the Cuttlefish by the subterranean search for water enacted by plants, this artwork takes him inland to a cemetery in Campbelltown in the centre of lutruwita/Tasmania where a rare and endangered species of wild orchid, known colloquially as the Graveside Leek Orchid, grows. De Carvahlo's work demonstrates ways to care for endlings like this orchid which may provide new ways to consider human and more-than-human engagements. de Carvahlo's work advances a relational approach between human and more-than-human to construct a multispecies approach to kinship in the Anthropocene.

In Chapter 8, "Conjuring up Ghost Species: On Photography and Extinction", Shane McCorristine, William M. Adams and Adam Searle turn to photography as a memorial technology. "Photographs provide a spectral view of absence, but also a truth-claim about presence" (x) they write, and when they depict extinct or near extinct species, photographs also have a special role in relation to conservationists and conservation politics. Photographs of extinct species (and they focus in particular on images of "endlings") create zones of liminality. These images destabilise the boundary between presence and absence, the visible and invisible, and the living and the dead. McCorristine, Adams, and Searle advocate for a nuanced approach to photography as critical to understanding the visual culture of extinction and the way that it is memorialised.

In Chapter 9, "An Elegy for an Ecotype: Eva Saulitis's *Into Great Silence* and the Extinction of the Chugach Transient Killer Whales", Jennifer Schell turns to literature as a site of memorial culture, examining marine biologist Saulitis' memoir which records, over time, the effects of the Exxon Valdez oil spill into Alaska's Prince William Sound and the impact of this on a community of killer whales belonging to the Chugach transient ecotype. Working with Deborah Bird Rose's notion of the "death zone" where the living and dying come into contact, and committed to a decolonial approach, Schell examines the depiction of Prince William Sound as a multispecies community in which human and whale cultures mingle. This chapter also examines the decolonial approach taken by Saulitis and the parallel constructed sensitively in the text between the dying killer whales and the impacts of colonialism on the indigenous community of the area, including on the loss of indigenous language. More broadly, Schell examines the literary history of elegy and the capacity of this form to facilitate engagement with loss and mourning without consolation or resolution. Engaging with elegiac reflections on species loss, and on the loss of individual animals, gives us a blueprint for living ethically in the death zone in the Anthropocene.

In Chapter 10, "Lost Species, Lost Worlds: Memorialising Extinction in the Art of Todd McGrain and Chris Jordan", I consider art as a memorial site for mourning lost species. The chapter brings together two artists who have produced work about birds, extinction, and endangerment: Todd McGrain's Lost Bird Project consisting of large bronze sculptures of extinct birds, placed and encountered in the location in the United States in which they were last seen in the wild and Chris Jordan's Midway Atoll series of photographs of decaying bird remains and the plastics they have imbibed. Working with art that memorialises and raises awareness about species loss and diminishment, this chapter considers the representation of extinct and endangered species and their worlds as they slowly fade from view. It examines art as offering an affective engagement with extinction politics and a complex nexus point for processing environmental grief, memory and forgetting, and human complicity in environmental destruction in the Anthropocene.

In the final chapter, "Psittacine Extinction Story: I Once Loved a Lorikeet", Hélène Frichot memorialises a Rainbow Lorikeet that enters her family's life during the COVID lockdowns imposed on Naarm/Melbourne (Australia). Layering the experience of her family in lockdown, with that of the Rainbow Lorikeet rehabilitating for fourteen weeks in her home, Frichot examines the collective trauma of "concatenating" environmental crises and the ongoing pandemic. Her cross-species entanglement with this bird, whom she calls Lori, prompts a reflection on the plight of the critically endangered Orange Bellied Parrot, and those conservationists working for their survival, and on the changing role of zoos as sites of research and conservation in the Anthropocene. Frichot concludes by turning to the work of Deleuze and Guattari and to Adrian Parr's work on memorial culture, to advance the idea of a constructive and future-focussed memorial culture of extinction.

Notes

1 See https://extinctionrebellion.uk/event/london-rebellion-extinction-march-there-is-strength-in-grief/

2 This responsibility is, of course, not evenly distributed amongst *all* humans. The impact of factors such as race, class, geographical location, economic position are central to understanding the differential responsibility for and experience of the impacts of the Anthropocene and associated phenomena such as climate crisis and extinction.

Works Cited

Brondizio E. S., J. Settele, S. Díaz, and H. T. Ngo, eds. *Global Assessment Report on Biodiversity and Ecosystem Services of the Intergovernmental Science-Policy Platform on Biodiversity and Ecosystem Services*. IPBES Secretariat, Bonn, Germany, 2019, 1148p. https://doi.org/10.5281/zenodo.3831673

Chakrabarty, Dipesh. "Postcolonial Studies and the Challenge of Climate Change." *New Literary History* 43:1 (2012): 1–18.

Chrulew, Matthew and Rick de Vos. "Stories of Unravelling and Reworlding." *Cultural Studies Review* 25.1 (Sept 2019): 23–28.

Doss, Erica. *Memorial Mania: Public Feeling in America*. Chicago and London: University of Chicago Press.

Enright, Kelly. "Exhibiting Extinction: Martha and the Monument, Two Modes of Remembering Nature." *Cultural Studies Review* 25.1 (Sept 2019): 154–171.

Extinction Rebellion. *London Rebellion Extinction March : There is Strength in Grief*. Accessed 12 October 2019. https://extinctionrebellion.uk/event/london-rebellion-extinction-march-there-is-strength-in-grief/

Grusin, Richard. "Introduction." *After Extinction*, edited by Richard Grusin. Minnesota University Press, Minneapolis, 2018, pp. vii–xix.

IUCN. *The IUCN Red List of Threatened Species*. Version 2022-1, Accessed 14 September 2022. https://www.iucnredlist.org

Jørgensen, Dolly. "After None: Memorialising Animal Species Extinction Through Monuments." *Animals Count: How Population Size Matters in Animal-Human Relations*, edited by Nancy Cushing and Jodi Frawley. Routledge, London, 2019, pp. 183–200.

Kolbert, Elizabeth. *The Sixth Extinction: An Unnatural History*. Bloomsbury, 2014.

O'Key, Dominic. "Why Look at Taxidermy Animals? Exhibiting, Curating and Mourning the Sixth Mass Extinction Event." *International Journal of Heritage Studies* 27.6 (2021): 635–653.

Rose, Deborah Bird. "In the Shadow of All This Death." *Animal Death*, edited by Jay Johnston and Fiona Probyn-Rapsey. Sydney University Press, Sydney, 2013, pp. 1–20.

1 Layering Loss

A Conversation with Lucienne Rickard

Keely Jobe

In September 2019, artist Lucienne Rickard began drawing on a blank sheet of paper in the Tasmanian Museum and Art Gallery (TMAG) in nipaluna/Hobart. The animal she was drawing, the Bramble Cay Melomys (*Melomys Rubicola*), was a small Australian rodent found on a single island in the Torres Strait, and it had only recently been declared extinct. The last sighting was in 2009, and its extinction was confirmed by the International Union for the Conservation of Nature (IUCN) in May 2015. The Bramble Cay Melomys' disappearance was particularly significant because it was identified as the first mammalian extinction to occur due to anthropogenic climate change. Efforts to save it had come too late, and what killed it in the end was rising sea levels. The Melomys' extinction was seen by scientists and conservationists as an indictment on both state and federal governments, whose lack of funding and urgency suggested a broader absence of ecological care. It was also seen as an indictment on the public whose value system didn't apparently extend to an "uncharismatic," island-dwelling rodent (Purtill).

When Lucienne had finished the drawing, with a handful of visitors watching, she took an eraser and rubbed the Bramble Cay Melomys out. This was the beginning of *Extinction Studies*, a durational performance supported by Detached Cultural Institution and given a home in the Link Building of TMAG, in which the artist would draw in great detail recently listed extinct plants and animals from the IUCN Red List, only to erase them when they were complete. Her art practice was not at that stage known for being overtly political, but a spate of documentaries and news reports focusing on the current global extinction crisis had spurred Rickard to consider her role as an artist and question how her practice might be put to good use. *Extinction Studies* was the response. "The plan was to use the same sheet of paper for the entire project, laying each species over the ghosts of those that came before" (Jobe and Rickard).

Three years after its launch, *Extinction Studies* continues at TMAG, focusing exclusively now on critically endangered Australian species. In those three years, a cultural shift has also been occurring in close proximity, a shift that questions the roles, productions and focal points of memorial

DOI: 10.4324/9781003315957-2

culture in Tasmania. This is a shift that critiques the way cultural sites and institutions "coerce our emotional and intellectual responses to the things they display" (Proctor 18).

In 2021, for example, a colonial statue located less than two hundred metres from the museum was the subject of a creative reinterpretation by Palawa and Pakana (Tasmanian Aboriginal) artists Julie Gough, Jillian Mundy, Allen Mansell, Roger Scholes and Greg Lehman. The bronze figure depicted William Crowther, Premier of Tasmania from 1878 to 1879, but more notoriously remembered for the covert dismemberment of Tasmanian Aboriginal man William Lanney's remains in the name of science. The *Crowther Reinterpreted* project – which variously saw Crowther's hands and head painted red to signify the body parts he removed, Crowther's face masked behind a latex version of Lanney's, and Crowther's entire figure concealed inside a packing crate – demonstrated that for those seeking to reify a culture or an identity, some actions must be cordoned off, relegated to a distant past, while for others, the past is only ever "terrifyingly and uncomfortably present" (Harper 14). It also asked the viewer to consider the kinds of narratives and lives we are made to celebrate and remember, and what kinds of omissions our remembrance relies on.

Also in 2021, TMAG formally acknowledged and apologised to the Tasmanian Aboriginal community, past and present, for "museum practices that have caused profound suffering for Tasmanian Aboriginal people and their community" (TMAG *Apology*), including, but not limited to, the mutilation, theft and trade of ancestral objects and human remains. In 2022, they opened a new exhibition, *taypani milaythina-tu: Return to Country,* which "presents creative work from 20 Tasmanian Aboriginal artists responding to relationships between community and Ancestral objects, particularly those held in institutions outside lutruwita/Tasmania" (TMAG *taypani*). The push to have ancestral objects returned (albeit on a two-year loan) from institutions including the British Museum, the Victoria and Albert Museum in London, the National Museum of Scotland, the University of Cambridge Museum of Archaeology and Anthropology, the World Museum in Liverpool and the Derbyshire Record Office (TMAG taypani) hints at a form of restitution that gives greater voice and agency to groups and cultures once targeted by institutional suppression, manipulation and abuse, though some rightly question the ethics of loaning ancestral items to the community from whom they were stolen (Aitken). The skewed colonial understanding of ownership is not lost on the artists involved in *taypani milaythina-tu*, as suggested by Pakana artist, curator, academic and Indigenous Fellow at the University of Tasmania, Zoe Rimmer, who states on the walls of the exhibition,

> If a museum's responsibility is to 'preserve' objects for future generations then surely museums have a responsibility to empower us, through the return of our Ancestral cultural material, to maintain and invigorate

> cultural practices for future generations. To truly preserve culture, we must have the opportunity to continue to create it.

Today, outside the museum, one might come across a group guided by Nunami Sculthorpe-Green, a young Palawa leader who's walking tour *takara nipaluna* (walking Hobart) follows the route taken in 1832 by the Aboriginal resistance, as they progressed to the old Government House to negotiate an end to the Black War. The walking tour "gives voice to the Palawa perspective and continued presence within this modern city, exposing the layers of history beneath the asphalt" (takara nipaluna). For the tour group, the streets of commercial Hobart become reactivated, contested sites of national identity and memory.

Examples of Palawa cultural expression and self-determinism are especially powerful when held against a false narrative of Tasmanian Aboriginal extinction that, despite continued Palawa presence, persists to this day (Lehman 161). Celebrating Palawa culture within, or in proximity to, TMAG can even be viewed as a pointed, subversive political act, as this is the very institution that publicly displayed the skeleton of Palawa woman Trukanini from 1904 to 1947, cementing her in the public imagination not only as the "last of her race" (Lehman 161) but scientific proof of a doomed population (Taylor 412). To utilise TMAG as a space of contemporary Palawa cultural expression and determinism is at once to quash the pervasive extinction narrative the institution helped create, and to place Aboriginal culture at the centre of state and national identity and memory.

Within this dynamic landscape of memorial culture, Lucienne Rickard has continued to draw, to pay homage to species on the brink, and those already vanished. Her project has undergone its own adaptations, moving from extinct to critically endangered subjects, from a global to a local lens and from a timeframe that saw one species drawn and erased every day to a deliberate slowing down, allowing for a greater degree of detail and investment. With no end date in sight and no indication that extinction rates are falling, Rickard plans to go on working with, against and in awareness of the institution's history, and that of the cultural landscape in which she performs.

Keely: Museums across the western world are currently experiencing a cultural and ethical reckoning. When decolonising efforts turn their attention on cultural institutions, they frequently expose traditional, imperial collection practices as theft, display methods as spectacle, and narrative styles as curated ontologies replete with gaps, silences and suppressed voices. That museums and cultural institutions must respond to these charges has become an unavoidable part of their day-to-day running, forcing them to rethink not only the kinds of practices and materials that constitute collections but the very role of museums as centres for knowledge. What is the significance of performing *Extinction Studies* within this context?

Lucienne: It's such an interesting time to be working in a museum. Sometimes it feels like this hallowed space, a great centre of culture and learning, and other times I'll be in a backroom having a bite, surrounded by these Frankenstein creatures that still need a lot of attention from the taxidermist and the pantomime of it all seems so amusing. It makes me think about what's holding these places up. Is it all just silicone and sculpted Styrofoam? Is that the stuff of cultural institutions?[1] Of course, it's a lot more than that, but I think that the Styrofoam at the centre is a useful way of thinking about museums now. It points to the ways in which museums go about telling their stories – how they construct them and shape them, and what's at the core of that narrative. What gets removed and what gets tacked on. I think the work of decolonising a museum's day-to-day model is bound to be fraught when their collection is made of and constituted by imperial beliefs and practices. Though somewhat belated, the museum I'm performing in seems to be putting this decolonising work more and more at the centre of its practice – preferencing voices and stories that have previously been suppressed, acknowledging and apologising for the harm the institution has caused First Nations groups and individuals, accepting continued mistrust from marginalised voices, and recognising the exploitative or violent ways they have treated both human and nonhuman lives. Apart from acknowledging disrespectful and violent practices, I imagine a part of this decolonising work would also involve critiquing rigid systems of categorisation and representation that fail to demonstrate the nuances of identity, ecology and cultural memory. The fact that TMAG is a museum and art gallery allows for a more vibrant atmosphere of interpretation. When art and science come together in one space, the audience are offered more than just a hard wall of facts; instead, a deeper level of engagement and exploration is encouraged. A project like *Extinction Studies*, which is performed alongside those stiff, labelled figures, and which asks the viewer to consider nonhuman life as complex, valuable, intertwined and agential, fits nicely within this critical framework.

Keely: Alice Proctor suggests that one of the chief functions of a museum is to "shape identity and memory" (9). How does *Extinction Studies* speak to this idea?

Lucienne: I'm really interested in the role that remembering and forgetting plays in identity formation. There are always aspects of national identity that we're keen to remember and celebrate, and other aspects we're more inclined to forget. I think this project speaks to elements of national identity that are deeply uncomfortable, and therefore not predominant in the broader national memory. We're happy to celebrate Australian biodiversity when we're talking about tourism, for example, but having the worst record of mammalian extinction in the world is something we're very willing to overlook when talking about agriculture or extractive industries. The inclusion of First Nations' voices is relatively common now in notions of national identity, but if those voices remind us of the continued harm colonisation inflicts, if those voices demand justice or compensation, a wilful forgetting is likely to occur. Here in Tasmania, the thylacine is a state symbol, one almost obsessively brandished with pride. But linking the extinction of that species to a programme of colonial eradication doesn't seem to be a part of the narrative. We're happy to forget that piece of the story.

Extinction Studies includes examples of extinction from all over the world, the only parameter being that the species I'm drawing has been recently lost, in the last twenty years. Interacting with visitors has become a big part of this project, it's something I encourage in the viewer, and there is one pattern of response that I have noticed throughout the project. Many visitors are comfortable judging the extinction records of foreign countries, and that judgement can be particularly reflexive in response to China and India, but if I ever counter with Australia's world-leading extinction record, or focus on an Australian species, that reflex swings in the opposite direction. Suddenly, extinction is out of our control, a global issue, a human issue, not a local issue. I suspect a big part of this defensive response has to do with the wavering legitimacy of a particular way of life that has, until recently, been celebrated in this country. Many of the typically Australian figures that stand in for a larger notion of Australianness, figures of iconic labour that we put so much energy and capital into – the farmer, the miner, the logger, the tradie[2] – are revealed as fragile constructs when held up against our high rates of extinction. This revelation is genuinely frightening for some people, and in a public cultural institution whose role is to define what it is to be Australian, I am often witness to that fear and anger.

Keely: You've been performing *Extinction Studies* on and off for three years now with no definite end date in sight. What does the element of duration bring to this work?

Lucienne: It's worth acknowledging that the element of duration is only possible because the extinction list is so long and continues to grow with every passing year, but it was the scale of the extinction crisis that sparked the idea in the first place, so it has been vital to the project from the beginning. When I first started *Extinction Studies*, the plan was to draw one species a day and erase the figure each afternoon, but it quickly became clear that it wasn't going to work, partly because I'm a very slow drawer, but mostly because I found that a necessary level of investment was missing from that hurried pace. I wanted to think of each work as a kind of vigil, and so I altered the process to include as much detail as possible in every drawing. Now, each drawing can take anywhere from two weeks to a few months to complete, and I consider that level of investment appropriate for the vigil of an extinct species. Spending three months on a drawing also makes for a very affective erasing, which can be over and done with in under five minutes. The durational aspect is also a tactical response to the drastically short attention span people often have regarding extinction and

Figure 1.1 Detail of drawing from *Extinction Studies.* Xerces Blue is rendered in the foreground with the erased Yangtze Giant Softshell Turtle hovering behind. Photo credit: Jesse Hunniford. Courtesy of Detached Cultural Organisation.

environmental issues. It seems that if we do care about this problem, that care is likely to fade after a few hours. I wanted a call to attention that outlasted the tendency to forget. Showing up every day for months on end to represent a single species was the approach I took (Figure 1.1).

Keely: That's a huge investment of time for something that gets erased in a matter of minutes. I imagine a sense of "waste" becomes central to the work, especially in the moment of erasure.

Lucienne: Yes, some viewers get really upset. For the audience, that's by far the most confronting element of *Extinction Studies*, the five or so minutes it takes to wipe a species from the page. I'm often met with really emotive language. I'm destroying the artwork. I'm ruining it. What I'm doing is sacrilege. It's wasteful. It's wrong. At the moment of erasure, I have noted a distinctly divided response: either people's level of investment escalates in that moment or the audience member will walk away, will give up on the artwork. I don't mind admitting that the division is quite gendered. Women are more likely to sit with the discomfort, and men are more likely to walk away. Countless visitors have suggested I keep the drawing and sell it for conservation funding, that I'm not achieving anything by erasing the work, and that's the most rewarding moment for me, that's when I know the project is really succeeding because it's that sense of waste, and that questioning of achievement, that challenges the very value system that led to the extinction crisis in the first place, that is, a western capitalist value system. Time is money in a western context, and withdrawing this highly detailed, time-intensive work from the market by erasing it feels like a slap in the face for many viewers.

When I drew the American butterfly *Xerces Blue*, I spent three months drawing every tiny scale on the wings. There were around 114,000. It was very laborious. When I finally erased the butterfly, there was a palpable sense of outrage from some audience members. It was one of the more powerful moments in the project. There were gasps and tears. It really did feel like a waste. But I believe a sense of waste is a valuable response to *Extinction Studies* because it's making the viewer feel something, whether it's loss, grief, regret, shame, defiance or refusal. That affective response is an achievement in my eyes because it's also an acknowledgement that the species represented did exist and it's not here anymore. The fact of its absence is hopefully being felt (Figure 1.2).

Figure 1.2 Crowd gathers around the completed drawing of Xerces Blue. Photo credit: Jesse Hunniford. Courtesy of Detached Cultural Organisation.

Keely: I find the most haunting element of this work is that the images of species you have previously erased hover behind the current subject like spectres. Those erased species then build up over time on the same piece of paper. When others have written about or commented on *Extinction Studies* (see Rawson), the word "palimpsest" continues to emerge as a term of viewer significance in regard to these spectres. And there is a similar word that you have applied to this work, which is "pentimento." What do these terms reveal about the project as a site of memorialisation?

Lucienne: Both of those terms are useful in demonstrating the play between absence and presence that's crucial to memorial culture. A palimpsest refers to a manuscript on which text has been overlaid on earlier writing, and it's a useful term for this project because of its original link to economy and resources. Manuscripts used to be made of animal skins, and they were expensive and hard to come by, so you had to conserve them, reuse them if possible. When *Extinction Studies* was in its planning stage, I was very concerned that this particular act of memorial also be materially and environmentally ethical – I didn't want to churn through reams of paper to get the point across – so to minimise resources, a single piece of paper and an eraser was decided on. In truth, I had no idea that the ghosts of species erased would hover on the page. The palimpsest was a happy accident that made the project what it is. Now those earlier species refuse to be forgotten, and that's entirely fitting because if the project can't afford a resource-heavy process (both financially and ethically), then it also means that I can't afford to forget the things I've drawn. The species won't let me!

"Pentimento" refers to the presence or emergence of earlier works or forms that have been painted over, but it's applicable to *Extinction Studies* because of its etymological link to "repentance." I like the idea that the layering of significance expressed in this work might provoke repentance or remorse, the kind that would move a person to atonement and even action. I've also been influenced by the way Jean-Michel Basquiat utilised "pentimento," not in the traditional sense but as a conceptual tool. Basquiat would paint words into his artworks only to cross them out, and he did this knowing that the omitted word would pull the focus, that the word was made heavy with meaning simply by being crossed out. This idea fits with *Extinction Studies* beautifully because it's the erasing that holds all the meaning. Often, I find visitors trying to decipher the ghostly remains, straining to figure out what shadowy forms lie beneath the current drawing. When I see those erased figures pulling focus, I think of Basquiat's pentimento.

Keely: You mentioned before about pushing back against the rigid systems of categorisation often present in museums. I think the cumulative layering exhibited in your work goes some way to describing the tangled nature of ecology and the resultant ripple effect of extinction. That it becomes harder and harder to identify individual species the more you engage with the work suggests to me a feeling of engulfment, a fitting affect when confronting mass extinction (Figure 1.3).

Figure 1.3 Lucienne Rickard erases Xerces Blue. Photo credit: Jesse Hunniford. Courtesy of Detached Cultural Organisation.

Lucienne: Yes, the edges are always the clearest. The centre, the point of most overlap and thus the most engagement, is the point of greatest wear and inscrutability. I agree that thinking of extinction in terms of individual figures or events is not particularly helpful at this stage. That's why I get a bit upset with efforts to bring back the thylacine because I don't detect a broader ethics of ecological care in these efforts; it feels as if the human is still at the centre. I can't help but ask, is this really for the thylacine's sake?

Maybe that's a good question for the memorial as a whole. Is a memorial really for the absent subject's sake, or is it meant for those left behind? I enjoy the look of complexity and entanglement that accumulation brings to *Extinction Studies*, but I'm particularly intrigued by the physical blurring and bleeding into that accumulation brings. The more I engage with this memorial, the more the paper wears, and the figures I'm asking people to remember blur into obscurity. The accrual of signifying acts has an erasing

Figure 1.4 Large sheet of paper reveals the overlapping images of erased species. Photo credit: Jesse Hunniford. Courtesy of Detached Cultural Organisation.

effect; in fact, the first four or five species I draw disappear completely under those layers of significance.

I'm reminded of the Port Arthur memorial, which is located at the site of a massacre in which thirty-five people died in Southern Tasmania. All the victim's names are cast in bronze at this memorial, but there are two names (those of two young sisters) that get greater attention, and that's apparent because their names have been buffed to a shine. It seems that visitors like to touch these names in particular, as if in acknowledgement of the victim's presence or absence, and those tactile accumulations have a wearing effect. The names shine, but they're also slowly being worn down by the build-up of public emotion. One day, they'll be illegible. This makes me think that the memorial is as much for those left behind as those who are gone (Figure 1.4).

Keely: That element of public emotion is why I view *Extinction Studies* as not only a hybridised work of drawing and durational performance but as an active, evolving memorial to extinction; a site and experience that's connected to deep emotions but one that also encourages a response in the viewer – those that are left behind and those who are witness to extinction. What kinds of responses are you encouraging, and what kinds of responses are you witnessing?

Lucienne: I don't mean to be a bully, but I want people to feel loss. I want people to feel a sense of responsibility and accountability. Those unpleasant, difficult emotions that we choose to switch off, I want to switch those on for the viewer. I'm gratified when people move between emotional states, from grief to anger, from helplessness to empowerment and from stunned inertia to inspired action, because witnessing mass extinction is alarming and it should provoke a range of responses. The really enjoyable element of this work though is when it inspires wonder and curiosity, not only for the species that are gone but for what we have left. I strongly advocate for more curiosity. I want to foster a sense of empathy and connection, and for people to see the similarities between themselves and whatever I'm drawing, especially those viewers inclined to think of humans as separate from or above the subject. The crayfish that passes the burrow down to his young demonstrates a practice of inheritance not unlike humans. The scales of a butterfly, which are made from a substance similar to human fingernails, and also a substance produced by mushrooms, show that we share chemical and material similarities with the nonhuman world. The Yangtze Giant Softshell Turtle was functionally extinct when I drew it. The last known female had recently died during the fifth attempt to artificially inseminate her. Though this animal's appearance was almost alien, women in particular reacted very strongly to the story of her unconsenting and violent end. They saw threats to bodily autonomy as a cross-species experience. I'm keen to drive any moment in which an understanding of kinship can be extended to nonhuman lives. If the viewer can think of the drawn subject as familiar, or even as kin, then they might also understand why that subject is worthy of memorialising.

Keely: The new iteration of *Extinction Studies* focuses exclusively on critically endangered Australian species. Why the change?

Lucienne: A big part of the change was about keeping my own curiosity alight. This kind of durational work can be quite numbing, and after the first sixteen months, I knew that I was flagging. I also suspect that the audience was becoming overwhelmed with hopelessness. I know I was. Critically endangered species are more relevant to a local audience – I sometimes felt the global approach was an easy way out for Australians. We're doing so poorly with our plants and animals, and I want local audiences to develop a sense of urgency. With almost two hundred species on the critically endangered list, it's a great reminder that Australia's biodiversity extends beyond the classics – the kangaroos, the koalas and the echidnas.

Keely: Have you noticed a different emotional response to subjects that are not yet extinct?

Lucienne: It's a bit early to tell, I've only done two erasings, though I have noticed more specific questions being asked about the subjects. This might have something to do with a reduced scale leading to more concentrated fields of care. The Orange-Bellied Parrot is a good example. There was a very powerful set of responses to that erasing, responses that seemed to me more personal, and I would venture, more productive, perhaps because the species is locally placed, is relatively well known and can still be saved. I hope that people left that erasing determined to take some kind of action, to make some material change in their lives, even if it was planting a native tree in their backyard. Maybe the case for thinking globally and acting locally can be applied to a memorial like *Extinction Studies*.

Notes

1 These are some of the common materials used by taxidermists in the construction of specimen mounts.
2 For more on iconic labour and extractive industry as related to colonial Australian identity and emotion, see Askland; Skilton.

Works Cited

Aitken, Sarah. "Stolen Tasmanian Aboriginal Artefacts are Finally Home. But There's a Catch: They're Only on Loan." *The Guardian*. Accessed 20 October 2022. https://www.theguardian.com/australia-news/2022/oct/19/stolen-tasmanian-aboriginal-artefacts-are-finally-home-but-theres-a-catch-theyre-only-on-loan

Apology to Tasmanian Aboriginal People. Tasmanian Museum and Art Gallery. Accessed 15 February 2021. https://www.tmag.tas.gov.au/about_us/apology_to_tasmanian_aboriginal_people

Askland, H. H. "Mining Voids: Extraction and Emotion at the Australian Coal Frontier." *Polar Record*. Accesssed January 2020. EBSCOhost, https://doi.org/10.1017/S0032247420000078

Harper, Andrew. "Crowther Reinterpreted." *Island Magazine*, vol. 164, 2022, pp. 10–17.

IUCN Red List of Threatened Species. Accessed 8 November 2022. https://www.iucnredlist.org/

Jobe, Keely, and Rickard, Lucienne. "The art of refusal." *The Living Archive; Extinction Stories from Oceania*. Accessed 5 May 2020. https://www.extinctionstories.org/2020/05/05/the-art-of-refusal/

Lehman, Greg. "Interest and Truth in History: A Review of Truganini: Journey through the Apocalypse." *Aboriginal History*, vol. 45, 2021, pp. 161–168. https://search.informit.org/doi/10.3316/informit.415435059609145

Procter, Alice. *The Whole Picture: The Colonial Story of the Art in Our Museums & Why We Need to Talk About It*. Cassell, London, 2020.

Purtill, James. "An Australian Rodent has Become the First Climate Change Mammal Extinction." *ABC Triple J Hack*. Accessed 20 February 2019. https://www.abc.net.au/triplej/programs/hack/bramble-cay-melomys-first-climate-change-mammal-extinction/10830080

Rawson, Jane. "Wildlife's Whispered Traces; Extinction Studies." *The Monthly*. Accessed 13 February 2020. https://www.themonthly.com.au/blog/jane-rawson/2020/13/2020/1581557758/wildlife-s-whispered-traces-extinction-studies

Rickard, Lucienne. *Extinction Studies*. Tasmanian Museum and Art Gallery, Tasmania, 2019.

Rimmer, Zoe. *taypani milaythina-tu*. Tasmanian Museum and Art Gallery, 2022–2023.

Skilton, Nick. "Mining, Masculinity, and Morality: Understanding the Australian National Imaginary Through Iconic Labor." *Gendering Nationalism: Intersections of Nation, Gender and Sexuality*, Palgrave Macmillan, 2018, pp. 31–47. https://doi.org/10.1007/978-3-319-76699-7.

takara nipaluna. "Theatre Royal." Accessed 8 November 2022. https://www.theatreroyal.com.au/shows/takara-nipaluna

Tasmanian Museum and Art Gallery. Accessed 8 November 2022. https://www.tmag.tas.gov.au/home

Taylor, Rebe. "Genocide, Extinction and Aboriginal Self-Determination in TasmanianHistoriography." *History Compass*, vol. 11, no. 6, June 2013, pp. 405–418. https://doi.org/10.1111/hic3.12062

taypani milaythina-tu: Return to Country. Tasmanian Museum and Art Gallery. Accessed 8 November 2022. https://www.tmag.tas.gov.au/whats_on/exhibitions/current_upcoming/info/taypani_milaythina-tu_return_to_country

2 Beginning and Endling

Archival Atmospheres, Extraction and the Case of Aotearoa New Zealand's First Rose-Crowned Fruit Dove, or Specimen OR.030538

Rosie Ibbotson

In April 2022, I travelled to Te Whanganui-a-Tara Wellington[1] to visit the dead body of a fruit dove. I had become aware of this bird in 2020, when – several months after his death in August 2019 – news was published of his short life, accompanied by some claustrophobic photographs taken during his last hours. The bird was a juvenile rose-crowned fruit dove (or *Ptilinopus regina*), a species for which the nearest known wild population is in Australia. This particular dove had flown thousands of kilometres to Aotearoa New Zealand, likely pushed by the strong westerly winds that had been blowing in the days around his arrival (Miskelly "First" 564), and had alighted on a ship stationed in an oil field in the South Taranaki Bight between the country's two main islands. Having been discovered by a concerned crew member, the dove was sent to the mainland on the next scheduled helicopter flight. Upon arrival, however, instead of being protected as a vagrant species, he was soon destroyed by Biosecurity New Zealand, and his body was later accessioned into the collections of Te Papa Tongarewa, Aotearoa's national museum (Miskelly "First" 564). Within days, therefore, this young bird had gone from becoming Aotearoa's first *Ptilinopus regina* – making this a new native species in the process, to being the last of a species now locally extinct in New Zealand, to becoming a seemingly peerless museum specimen.

This recent and confronting chapter in Aotearoa's extensive history of anthropogenic avian extinctions – which has disappeared (among others) the huia, the koreke (New Zealand quail), and the whēkau (laughing owl) – is the starting point from which the themes of this essay unspool. While the *Ptilinopus regina* is not globally extinct, the current status of its extant population – or at least, its status when last assessed in 2016 – is simultaneously described by the International Union for Conservation of Nature as being of "least concern" and as "decreasing" (IUCN) – a dismaying juxtaposition.[2] Furthermore, the disquieting story of this species in New Zealand raises critical questions about museums, borders, and belonging,

DOI: 10.4324/9781003315957-3

and the geopolitical conditions under which species loss and related forms of extractive environmental violence occur. Through both his charismatic biography and its implications for his species more broadly, the particular bird now described as "specimen OR.030538" seems to invite consideration of museological framings of extinct animals, and what these contexts might mean for practices of mourning and memorialisation. While museums are ostensibly a key infrastructure of social memory, the extent to which they serve as sites of mourning and memorialisation is obscure.

By thinking with OR.030538, this chapter seeks to explore how museological practices and their representational optics can blur boundaries between the individual and the species; the visible and the invisible; the anthropogenic and the "naturogenic"; and the living and the dead. In considering the curious ontology of Aotearoa's first and (currently) last native rose-crowned fruit dove – as a museum specimen that does not just represent but arguably *is* the entirety of a local species; as a supposedly representative museum specimen that was instead "immortalised" in an otherwise fleeting state of juvenile plumage; as a local extinction that was both deliberate and, unusually, directly witnessed by humans – the gaps, erasures, and illusions in the museological records and narratives of other species become more starkly apparent, as do the ways in which technologies of archiving and display consciously or otherwise circumvent these absences. This chapter therefore also reflects on how historical museological practices have exhibited complicity with extinctions, and the extent to which archival forms of memorialising loss might seem to normalise rather than mourn it. Crucially, this chapter aims to highlight more broadly the ecological agency of museums, and the manifold entanglements of environments and archives.

Despite the fruit dove's unusual case, OR.030538 is not on display, and visiting him requires going into the Te Papa stores.[3] Signing in to the cavernous room in which he is kept, I learned that I was the first visitor to request to see him, which surprised me given the newsworthy nature of his story. This realisation was also moving, and it had the effect of alerting me to another aspect of my visit: while I had been seeing this as a research trip, it was also an act of mourning. I briefly pondered the extent to which these two things are recognised as compatible. Learning of his lack of visitors – in addition to what I already knew about his lonely journey – added urgency to my impulse to bear witness to his existence, and compounded my sense of the dove's isolation, which was underscored further still by his location near one of the far corners of the store. This is arranged largely taxonomically, though outsized taxidermy birds on mounts perch here and there among the forest of cabinets. Walking through the steady, calibrated cold of the archival space to reach the cabinet in which he is housed, I was aware of feeling inadequately prepared somehow for the moment of finally seeing the dove – something that I later saw reflected in the clumsy quality of the photographs I took during my visit, which came out as yellowy, unmethodical, and abashed.

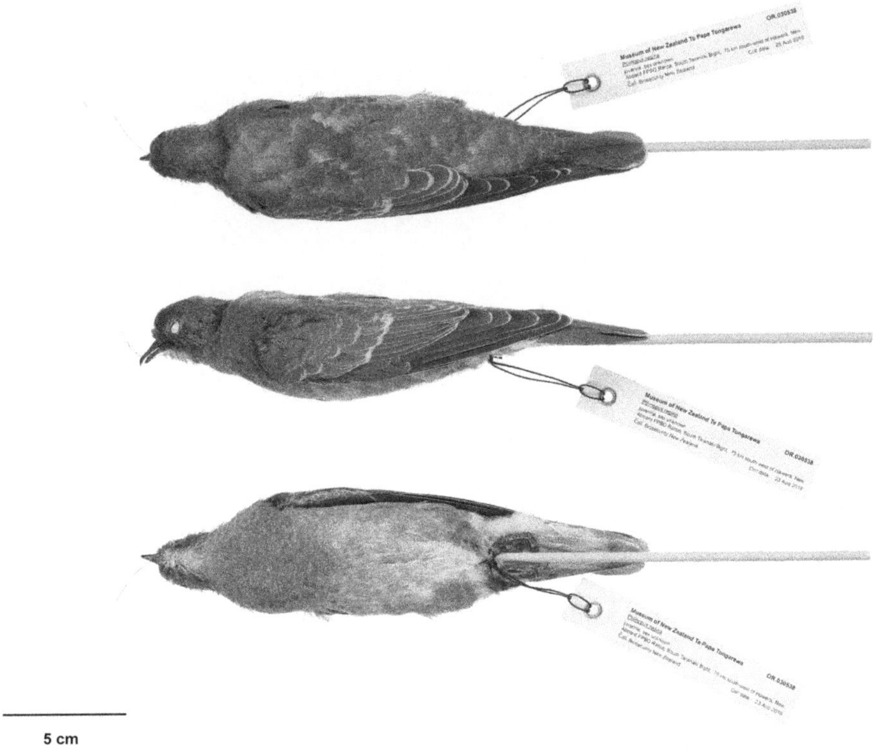

Figure 2.1 Te Papa Tongarewa specimen OR.030538 (*Ptilinopus regina*, collected 23 August 2019, aboard FPSO Raroa, South Taranaki Bight, 75 km south-west of Hawera, New Zealand. CC BY 4.0).

Looking back, at this stage I had not yet accepted referring to the fruit dove by the name I now use to think about him – Ori. I had let my eyes skate past his stubbornly unmemorable accession number, and when jotting down stray thoughts and notes to myself, I had tried to reproduce the six digits that follow "OR." as faithfully as time allowed. But even when faced with the fastidiously reproduced accession number among these otherwise provisional musings, only the letters sounded in my head, and the dot of the full stop soon morphed into an "i", reminding me of the name of a friend overseas who I have not seen in years but miss. After my visit to see the dove in the drawer, however, I appear to have unconsciously given myself permission to embrace the name – not as something which was "of" the fruit dove or seeking to be imposed more broadly, but as a (then) private acknowledgement of a connection perhaps forged by our meeting. Later, when writing this chapter without using my name for the dove was proving more difficult than I had anticipated, I looked for the existence of this word as a term or a name in other languages, and reflected on the multitudes suggested by the

Figure 2.2 OR.030538 (identifiable by the attached stick) in the storage drawer. Author photograph.

many different definitions, concepts, and even acronyms found. In ways that seem vividly consonant with elements of the fruit dove's story, Te Aka, the Māori Dictionary with the small green bird on its website, notes that "ori" has a range of meanings, including "strong wind", "victim, prey", and "contaminated place". This green bird cartoon, I also notice, is animated when attention is focused on it through the hovering of a cursor.

Before visiting Ori, I had pictured his body as somehow alone, perhaps separate from others within his storage drawer. This sense derived from his epistemic singularity, but also perhaps from a tripartite photographic image of his body which accompanies the entry for OR.030538 in the Te Papa online catalogue. This shows him as a lifeless, elongated "study skin" – mounted on a stick, missing one wing, and with a label attached by a cord to his bound feet – and stacks views of his back, profile, and front floating within white, empty, clinical-seeming space (Figure 2.1). However, as the draw opened with a pronounced waft of camphor, I saw that he was in fact nestled snugly among others of the same species (Figure 2.2). This realisation greeted me with an initial rush of misplaced relief, before my thoughts returned to the ecologically deleterious contexts under which museum collections were initially built up in Aotearoa, and the potential violence within the individual circumstances through which the body of each bird had come to be there. Alongside the other *Ptilinopus* specimens, all reclining in the spatially expedient but anatomically improbable manner of study skins

Figure 2.3 OR.030538 (identifiable by the attached stick) among other Ptilinopus regina skins. Author photograph.

within the different quadrants of the drawer, the dove's juvenile plumage was more obvious (Figure 2.3), making the burden of representation that his museological context implies all the more fraught.[4] He was not a fungible member of a homogenous group – and, regardless of their particular biographies, neither were the other birds in the drawer. Indeed, it is interesting to consider the relationship between OR.030538 and the older *Ptilinopus regina* specimens in the Te Papa collection, and how he might be seen to have reconstituted their position within the institution. As a state museum, which today emphasises showing the nation to the world rather than showing the world to the nation, the arrival of the vagrant juvenile fruit dove as a new species for Aotearoa had recast the other *Ptilinopus regina* specimens as no longer "exotic". The very bird that was executed as a supposed outsider to New Zealand was the one that now lent a certain "belonging" to this species within the Te Papa collection. I noticed that the "*Ptilinopus regina*" label on the archival drawer was still black, a convention in this store that denotes non-native birds, rather than red, indicating New Zealand species. In imagining this label being swapped for a red one in line with the rose-crowned fruit dove's "graduation" to the status of a native bird, I wondered how many other labels might light up red in the future, as climate change continues to effect shifts in birds' behaviours, distributions, and migratory patterns (Davis and Watson 239).

Adding to Ori's ontological complexity is the division of his remains into four different parts, spanning three different areas of the Te Papa collection. The study skin is only part of the bird, although another piece of his body – the detached right wing (OR.030538/1) – remains nearby in the drawer. While this arrangement – sometimes called a "spread wing preparation" – is a relatively recent convention of bird skin taxidermy (Webster 4; Claramunt and Wright 134), designed to minimise handling of the main body during data gathering specifically involving the wing, it is nonetheless confronting to see this dismemberment. Observing the wing held once again besides the dove's body (Figure 2.4), I was reminded of Donald Preziosi's notion of museological "*re*-member[ing]", which illuminates the inherently fragmentary nature of the collections held by museums – and indeed sometimes of the objects themselves. In drawing attention to how museums "re-member", Preziosi questions the narrative implications of objects being reassembled into seemingly comprehensive or representative histories (101) – something that in the case of animal specimens is fundamentally premised upon "species-thinking" and the erasure of their individuality (Stark 68; Chrulew 141). While less visibly absent from the study skin than the wing, the dove's tissues reside elsewhere in the museum's storage – in the spirits collection – while some of his bones are stored with other skeletal matter.

The distributed body of the dove adds to a sense of his liminality, a theme that more broadly characterises the known biography of his short life. As a vagrant bird, he crossed various forms of border and moved between different zones – geopolitical, climatic, temporal, as well as cultural and conceptual ones. Indeed, vagrant individuals – those "observed outside [the species'] normal geographic range" (Callaghan et al. 72) – challenge the common myth that all members of a given species act alike, and by extension are largely interchangeable.[5] Perhaps the most fateful boundaries for his case, however, were the discordant legal ones in which he found himself enmeshed, and how these were interpreted. Colin Miskelly, the ornithologist who identified the bird as a *Ptilinopus regina* and has documented his story and relationship to Aotearoa, has described how – despite the existence of New Zealand law "that provides automatic protection to vagrant birds" – the

Figure 2.4 OR.030538 and OR.030538/1 held together. Author photograph.

bird's death "arose partly due to misalignment of two different pieces of legislation" ("First" 566):

> Since 1996, the Wildlife Act 1953 has covered waters out to the edge of the New Zealand exclusive economic zone (200 nautical miles = 370.4 km from the coast; Miskelly 2016). In contrast, the Biosecurity Act covers only territorial seas (12 nautical miles = 22.2 km from the coast). As the bird was flown to shore by helicopter from a vessel anchored *c.* 73 km offshore, Biosecurity New Zealand staff treated it as an importation of a live bird, which is not covered by an existing Import Health Standard...
>
> ("First" 566)

It therefore seems possible to imagine various scenarios in which things turned out differently for the fruit dove. As Miskelly has noted, this is

> [o]ne of the ironies in this case... if the dove had flown two degrees either north or south of its chosen course, it could have flown a similar distance and made landfall in Taranaki or Golden Bay. In all likelihood it would have disappeared into the treetops...
>
> ("A new")

The idea of this now-defunct possibility is reinforced by the relative regularity with which other self-introduced birds join the ranks of Aotearoa's native species. If Ori had instead alighted on the landmass, his story might have resembled that of the Bulwer's petrel – another vagrant bird who made his species native to New Zealand, died as the local endling, and subsequently was collected by Te Papa and turned into specimen OR.025739. (Affectingly, the beach in Horowhenua on which this petrel's body was found looks out towards the oil field where Ori landed.) Instead, however, the arrival of the *Ptilinopus regina* became "the first reported occasion where a new vagrant bird species to New Zealand has been intercepted at the border and killed due to biosecurity concerns" (Miskelly "First" 566). Ori's case therefore highlights the constructed and confused nature of borders, and demonstrates how animals as well as humans are subjected to their violence.

It also seems possible to imagine a scenario in which the fruit dove – or his body – did not end up in a museum collection. For all that it can be confronting to see animals' bodies in archival storage, it is unsettling to imagine local or global extinctions happening unnoticed and undocumented, without the opportunities for ongoing research, contemplation, and memorialisation that museums can foster. However, the extent to which museums currently facilitate these practices can vary. Especially within the resting place of a museum – in particular one that is an institution of the state – Ori's fragmented body reflects his hermeneutical complexity, and might be seen as a site on which various conflicting themes converge. He was the first of his species to become native to Aotearoa, but soon afterwards became the last of his kind there. In travelling the vast distance to New Zealand waters, he took an atypical path when compared to his peers and relations, and yet within a museological context he becomes framed as somehow "representative" of his species more broadly. He was a precocious individual too, not only in making his anomalous journey while still a juvenile developing his principal flight feathers, but also in terms of his early birth: as Miskelly has described, clues within the bird's plumage suggest him to have been less than ten weeks old, presumably having "hatched from an egg laid in June or July" which resulted "from an unusually early breeding event" ("First" 566). (Unlike some other species within the Columbidae family of birds, the *Ptilinopus regina* is capable of fledging while still some way off their full size.) Ori's strange and lonely journey has led to the reclassification of his whole species: the *Ptilinopus regina* is now classed as native to New Zealand (Miskelly et al. 259). He is a celebrated new arrival, but was killed by the state. The premeditated nature of his death means that he occupies an unusual place within histories of anthropogenic extinctions: the very moment of the rose-crowed fruit dove's local extinction in Aotearoa was witnessed by, or at least in the presence of, humans.[6] The dove lies motionless in the archival drawer, but poignantly, this is as a result of his life of extraordinary movement.

Ori's biography therefore provides a particularly striking illustration of what can get lost behind the objectification of methodically generated accession numbers, and within typical museological taxonomic logics. While he arguably joins the captivating community of animals who, as Samuel Alberti has put it, "refuse to be constrained by their zoological classification" (1), he perhaps also illuminates the anthropocentricity inherent in the idea that this resistance to human knowing applies only to some "exceptional" creatures. Rather, in the vivid story of the dove's arrival to Aotearoa, an agency is clearly apparent which makes him decidedly un-specimen-like, and he defies classification even as he is repositioned within this, and transforms it for other rose-crowned fruit doves. However, as André Krebber and Mieke Roscher have observed as part of their explorations of "animal biography", "history is dominated by attempts that try to standardize, deindividualize and automatize the behavior of animals" (1), and Ori's placing among the bodies of his species-mates indeed seems to fold him into a depersonalised museological inventory. This somewhat obscures his affect, and shrouds the mechanisms and processes by which these particular deaths have occurred – and by extension, how larger losses within and of species have come about. Yet it is in the particularities of his own biography that Ori invites the consideration of the others as individuals who lived unique lives, most likely separated by substantial time and geographical space until they entered the museum collection and were forced together within the suspended animation of the archive. While Ori's charismatic backstory might be seen as exceptional, it could also highlight the gaps in what is known about the individual lives of other specimens, as well as the erasures in narratives, knowledges, and museological records that occur when fragments are conjured into the illusion of a generalised, Platonic whole.

Indeed, the fact that museums and related institutional forms positioned individuals as placeholders for species seems to have contributed substantially to extinction. In the nineteenth century, when the conjoined phenomena of museums and European imperialism were spreading across the world, the bodies and skins of birds and other animals were traded around expanding global networks, and were frequently harvested en masse for this very purpose. At the time, museum collections often pursued the conflicting aims of being encyclopaedic and representative yet exceptional and high quality, and this led to organic ecological assemblages being fragmented and depleted in their conversion into anthropogenic, ostensibly controllable archives. The desire of museum directors to populate their arks meant that not only were endemic species in high demand as specimens both within and far beyond their regions of origin, but the slaughtering of animals and the exportation of their preserved bodies also became a means by which all sorts of collection items deemed *de rigueur* might in exchange be sent in the other direction. In nineteenth-century Aotearoa, for instance, material culture from the so-called "old world" was highly sought-after by colonial museums. This is perhaps exemplified by a report published in 1894 by the *Otago Witness*, describing a meeting of the Otago branch of the New Zealand Institute in which it was claimed that "No museum was complete without an Egyptian mummy" (21). Objects like this were frequently paid for in the lives of native birds, whose numbers sometimes dropped precipitously as a result, and whose environments were devastated by hunting practices and related forms of colonial violence and extraction. Material culture in New Zealand museum collections that was imported and accessioned in the long nineteenth century, then, is sometimes indexical of catastrophic avian destruction.

Historically, therefore, museums must be understood as agents of extinction, both directly through some of their acquisitions and exchanges, and more generally in terms of their profound entanglements with the proliferation of colonialism. While the conditions under which this vagrant dove died and was acquired by Te Papa of course differ substantially from those of birds that entered museum collections in the long nineteenth century, histories of colonial and museological environmental violence are inextricably connected with the context that now frames his body. So too are colonial histories of species introduction or "acclimatisation", the ongoing effects of which have contributed to New Zealand's approach to biosecurity. The "ecological imperialism", to borrow Alfred Crosby's term, that saw the widespread introduction of a curated selection of non-native plants and animals to Aotearoa has given rise to a number of the primary industries that, among other things, biosecurity practices and eradication programmes are seeking to protect, and it has also precipitated numerous cautionary tales about allowing locally novel biological agents – including species now considered "invasive" – to become established within this archipelago's fragile ecosystems. Ori's story therefore highlights the ongoing nature of colonial environmental violence, and the role imagined borders are made to play in this. Indeed, it is almost possible to read into his journey a kind of subversion of the logics of colonial acclimatisation practices: instead of being forcibly imported, he self-introduced to New Zealand waters, and unlike so many of the transported plants and animals who were to die at sea, he survived his journey, only to be killed upon landing in Aotearoa.

Indeed, the extractive logics at work in both historical museum practices and colonialism more generally resonate with the context of the fossil fuel industry, which played such a crucial role in Ori's story. The ship on which he landed was the *FPSO Raroa*, a tanker converted into an oil storage and processing unit, and now fixed in place in the Maari oil field via a permanent internal mooring. While different in fundamental ways, including in their relative scales and temporalities, both fossil fuel industries and many colonial museums (including in New Zealand) have exhibited complicity with extinctions – both in leveraging the remains of disappeared species (as converted into crude oil or specimens) and in contributing to the extinction of others. This is of course not to suggest equal culpability, nor is it to imply that today's curators are not keenly aware of the fraught histories of their institutions. However, thematic parallels of these industries also extend to the way in which they have projected damaging and Eurocentric ideologies which have sought to position humans as above, and somehow separate from, surrounding ecologies. In visiting the dove, my own implication within these destructive processes – and the illusory nature of distance from a subject of study – was thrown into sharp relief, as I reflected on my dismally ironic decision to fly to Wellington, over sea not far from where the fruit dove had landed, on alloy wings powered by fossil fuels. What species, I wondered – which myriad individual animals and bodies – had over deep geological time become the crude oil that made the jet fuel now being combusted on this very plane? Which species – and with what consequences for individual animals – were pushed closer to extinction by this flight and the myriad others taking place around the world that day – that month, this year? I was reminded of Timothy Morton's notion of "dark ecology" and its aesthetics, which he describes as reflecting the structure of a noir film: "The noir narrator begins investigating a supposedly external situation, from a supposedly neutral point of view, only to discover that she or he is implicated in it" (16–17).

The fruit dove's arrival on the stationary vessel also resonated with certain cultural narratives, or at least corrupted versions of them. A particularly striking – though subverted – echo is with the dove in the story of Noah's Ark, whose release from the ship and subsequent return with an olive branch indicated the presence of a safe place to land, and by extension the success of Noah's mission in weathering the ravages of this particular biblical environmental apocalypse. In this story, the archival vessel secures future biodiversity, and animals proliferate after the ecological crisis has abated. In the case of New Zealand's rose-crowned fruit dove, however, the ship is emblematic of anthropogenic environmental damage, and the dove's encounter with land is the opposite of an optimistic new flourishing for his species. Indeed, it is interesting to consider the killing of this juvenile *Ptilinopus regina* in the context of the substantial symbolism of doves. Across a range of different cultures and religions, doves have longstanding associations with hope, love, peace, and freedom (see Allen). The killing of this dove therefore might be seen to take on an especially heavy significance, and seems an apt metaphor for the anthropogenic violence fast and slow that has so devastatingly disappeared species – often out of sight, moored beyond the reach of sustained human attention, and in hazy and liminal zones which can all too easily fall through gaps in oversight, answerability, and understandings of culpability.

While Ori's journey was remarkable, it should be noted that his arrival on an offshore oil platform is not an uncommon phenomenon among vagrant and migrating birds. Indeed, there is a growing literature on what Robert Ronconi et al. describe as "deleterious bird-platform interaction" (34), which seems to be exacerbated by the particular lighting used on many rigs, as this has the effect of disorienting travelling birds, frequently contributing to their deaths (Mosher). Semi-regular news stories describing birds landing on these covert anthropogenic archipelagos have in the last few years included numerous owls, a water rail, an osprey named Carla by the crew of the ship she landed on, and Pedro the pigeon (see McKim; "North Sea oil"; "Off-course migratory bird"; "Bird airlifted"; "Osprey hurt"; "Lost racing pigeon"). The news articles highlighting these particular birds' journeys – like Ori's – attest to the dedication of many rig workers in caring for these birds, before their relocation to rehabilitation facilities on land. In the case of Chopper – a migrating short-eared owl found on a North Sea gas platform, and named as a result of his helicopter evacuation to the British mainland – one of the rig workers who rescued the owl planned a visit to see him at the rehabilitation centre before his release back into the wild (Donnelly).[7] More similar to Ori's case was the 1989 sighting of a lone Nicobar pigeon on an oil rig in the Timor Sea. The platform in question was not far from mainland Australia, where another individual of the same species (*Caloenas nicobarica*) was to self-introduce in 2017. This bird too was intercepted by biosecurity forces, but instead of being killed was permanently detained in Adelaide Zoo (Davis and Watson 238).

The ecological malevolence of fossil fuel rigs alongside their occasional portrayal as sites of temporary shelter for birds echoes a conflict between differing conceptualisations of museums. The environmental agency of archival institutions, and their entanglements with climate change – including in some cases through their acceptance of sponsorship from fossil fuel companies (see, for example, Evans) – mean that it is particularly ironic that museums in the Anthropocene are increasingly being positioned as refuges for collections on a volatile planet. The more pressing the environmental threats – the same ones that both contribute to and sometimes stem from extinctions – the more that museums are tasked with fortifying themselves against these contingencies, and the more they are presented as oases of stability within a precarious and unpredictable world. This feedback loop is ambivalent, as while it frames archival space as chambers that might keep out climate change, significant resources are expended on attempting to micromanage internal climates in a way that seeks to control both what is within the archive, and what is excluded or extinguished from it. The archive is thus a highly controlled, artificial environment that is defined by its borders, which form a pleural membrane between objects and the apparently "invasive" actants and pollutants of the outside world. In the context of encroaching climatological disaster, it appears as a Wardian case aboard a nineteenth-century sailing ship, or an oxygen mask as the air drains out of the cabin. Archival spaces thus both respond to and normalise – but are also perhaps in denial of – what Nathan Hensley and Philip Steer have described as the "generalized death drive of western life" (1). Indeed, preventative conservation practices within museum storage – which range from relatively straightforward-seeming interventions, such as the presence of camphor (see Shen),[8] to wider use of fumigants, to the installation of costly systems relying on highly specialised engineering – are frequently based on perpetually attempting to arrest or at least suppress the biological and inorganic processes that constitute the ecologies of the archive. Paradoxically, these practices are frequently energy-intensive, but measure their success through achieving an ostensible state of atmospheric stasis within the archival lung. It is therefore possible to interpret the museological archive as the fiction of a site where ecologies cease to exist. While it is in fact an environment in its own right (and often a mercurial one at that, as curators and conservators experience frequently in the course of their work), in positioning itself as an inert "non-ecology", where materiality might stand still, archival space arguably disavows lively entanglements within its boundaries, but remains on high alert for those on the outside. This mirrors the fruit dove's arrival to New Zealand and his treatment by the state: he presented the possibility of external agency within the curated ecologies of the archipelago, and was thus killed in what appears as a pre-emptive attempt to ensure this specious environmental stasis. Indeed, in a discussion highlighting the living ecologies within archives, Dani Stuchel has noted how unbidden "animal, plant, and abiotic changes which work against projects of human history are seen as failures, infestations, or disasters – they can never be properly archival" (1). Ori's becoming "archival" in this sense, then, was conceptually as well as logistically conditional upon his death.

The optics of archives also lend themselves well to extractivist metaphors, and researchers often talk of mining, excavating, and "digging things out". Indeed, Stuchel's observation that archives are frequently imagined as an "extractable resource" (1) highlights not just the susceptibility of their contents to instrumentalisation, but also how collections in the Anthropocene might be more vulnerable than they appear. These themes all converge in the case of DeepStore, in which the positioning of the archive as a climatologically stable bubble intersects explicitly with histories of capitalist extraction. Established in 1998, this is a vast storage facility used by museums, among other institutions, and located within the Winsford Rock Salt Mine (opened in Cheshire in 1844), the largest working salt mine in Britain. In it, the minerals from which the cavernous subterranean complex is terraformed engender an ostensibly controlled atmosphere, with the sort of consistent temperature and humidity conducive to preserving heritage collections (salt, of course, has a long history as an agent of preservation – including in early techniques of taxidermy (see Poliquin 25, 30)). However, DeepStore's expedient union of mining and archival space can be interpreted as haunted by environmental anxiety: the sense that valuable heritage objects need to be interred in a vault deep underground to give them their best chance of long-term survival seems somewhat ironic in the context of the Winsford Mine's continued operations (albeit not in the extraction of fossil fuels), including in the use of an area accessible via one of its shafts – number four, known as Minosus and run by the transnational giant Veolia – for the storage of hazardous waste ("Winsford Salt Mine"). These all-too-predictable juxtapositions recall literature scholar Heidi Scott's sobering observation that landfills "more accurately reflect our values than museums do; they are the cultural repositories of the Anthropocene, hermetically sealed to shield us from the reflection of our own profligacy" (589).[9]

The idea of archival material being stored within a bunker-type space that might not be as stable as it would initially seem is also relevant to the Svalbard Global Seed Vault. Established with the intention of protecting biodiversity and mitigating threats to food security, this facility is sometimes described as "the doomsday vault", and tunnels underneath a hill in the hopes of creating a secure environment for the storage of seeds from plant species from across the world. However, the unstable future that this project imagines is already here, as war has already forced at least one country to access the seed reserves (Tuysuz and Damon), and the facility itself has started to flood due to unprecedented ice melt (Carrington). As Tracey Heatherington has pointed out, its design – "like a military bunker and guarded like a bank vault... normalizes the fortification of seed commodities, even as it fails to protect agricultural communities from crisis events or the banal indirect violence of a harshly neoliberal global economy" (53). There is a connection, Heatherington seems to be arguing, between certain archival impulses, the security state, and the reluctance to confront the root causes of anthropogenic environmental violence. Indeed, this reluctance is arguably assisted by the idea of this archive as an "insurance policy", but such a view reifies the species-thinking of the archive and overlooks the wider, entangled picture – including all of the systemic environmental factors (and indeed cultural ones) required for different seeds to germinate and plants to grow and sustain other life forms. Furthermore, in prioritising security and boundaries over interaction with and animation by visitors, the potential of archives to be sites of memorialisation is foreclosed, and with this the opportunity to cultivate a broader ethics of care which might better engage audiences and communities with questions of their agency within planetary futures.

At times, the security of institutions' archival spaces appears to conflict not just with the involvement of visitors, but with environments for (more than) human animals more generally. One of the more striking ways in which this occurs is through the installation of gas-oriented fire suppression systems, which reify the idea of the archive as a bubble or lung. These systems respond to the threat of fire by piping in gas which changes the composition of the archival atmosphere. Ironically, however, a gas that was previously commonly used within these systems – bromotrifluoromethane (CF_3Br) – has been found to be deleterious to the ozone layer, highlighting further potential facets of the environmental agency of archives. While many fire suppression systems now use pentafluoroethane (CHF_2CF_3), often described in marketing materials as a "clean agent", this gas also has implications for the climate, and was classified in the 2016 Kigali Amendment to the UN's Montreal Protocol as a controlled substance, which had a "100-year global warming potential" of 3,500 times that of carbon dioxide (Ozone Secretariat 925). Thinking of these suffocating statistics in relation to museological space brings both the image of the planet and its atmosphere and the study skin into relation: when living birds are earmarked for taxidermy or archival preservation, the method through which they are killed often involves compression to the thorax, extinguishing their lungs of breath.

In addition to their potential environmental implications, these fire suppression systems and their aesthetics of the archival bubble appear – perhaps unusually for a mechanism of museological storage – to have gained some purchase within public imaginaries of collections and their institutional architecture. For example, a popular myth lingers around the iconic structure of Yale's Beinecke Rare Book and Manuscript Library, which claims that a system exists that would drain the oxygen out of the stacks in the event of a fire, and carries the implication that collections were prioritised over humans. While untrue, and in fact based upon a misunderstanding of how the Beinecke's reportedly pentafluoroethane-based fire suppression system works (Wagner), the persistence of this rumour is suggested by the existence of numerous articles from a range of different outlets that have set out to debunk this myth (see, for example, Wagner; "Fact check"). That this false rumour appears to be so compelling perhaps speaks to how museum visitors conceptualise their place – or lack of it – within archival environments. Similarly, research has suggested that there are substantial associations for some people between museums and death (Merriman 155), something that has no doubt been enhanced by the tomb-like monumental architecture of the archetypal Eurocentric museum. Furthermore, Rachel Poliquin's image of the taxidermy archive as "the breathless zoo" signals how the associations between museums and death might be particularly acute in contexts involving preserved animal bodies, and this reference to breathlessness also underscores a particularity of museum storage atmospheres – there is something in, or missing from, the air.

While Ori himself might be lifeless, it is interesting to consider the lively ecologies that accumulated on his body as he moved within the air of the outside world. What unique environmental archive do his feathers carry, and what might these record about his previous habitat, his social life, his journey, or even the Anthropocene itself? What museological stowaways and particulate matter are part of OR.030538 – maybe pollens, (micro)organisms, other genetic material, soot, atmospheric pollutants, microplastics, traces of crude oil?[10] And how does Ori's corporeal archive differ from those on the bodies of his species-mates within the storage draw? The clues that these furtive assemblages potentially contain about more-than-human biographies might – like Ori himself did – add to the narrative of his entangled species, and yet within the anthropocentric structure of the archive, they often remain overlooked by museological canons of attention. It is interesting – but perhaps inappropriate – to wonder how the enmeshed community within the storage drawer would be perceived within the Umwelt of a living *Ptilinopus regina*. However, even with more questions than answers, in recognising museum objects as unconscious archives themselves, the impossibility of policing boundaries between the inside and the outside of collections spaces is underscored.

Indeed, in various cultural imaginaries, birds represent a similar point about how boundaries might be transcended, and freedom more generally. Birds are, for example, a recurring theme in the work of Behrouz Boochani, a writer, filmmaker, and journalist who has drawn significant attention to the violence of borders, and has shared his first-hand experiences of the painful years he and many others spent detained in a refugee camp on Manus Island. Because birds suggest the possibility of living in innocence from geopolitical thresholds, it is therefore poignantly ironic that Ori was stopped at a border, and perceived as a potential "contaminant" or invasive threat to the nation he was arriving in. Further compounding this irony, the surveillant anxieties that precipitated his destruction have their analogues within the archival context he is in, where his lifeless remains are now themselves insulated against "invasive" agents. Once subjected to violent exclusion, Ori – or at least his body – is now deemed valuable enough to be included and protected. That his particular resting place is in a national museum speaks to his significance to Aotearoa and perhaps a posthumous belonging, even if a different arm of the state, Biosecurity New Zealand, did not see his life in this way.

Indeed, Davis and Watson argue that vagrant birds should be seen as "climate refugees", and also note the crucial role they play "as first responders to environmental change" (238). When read in relation to Ori's case, this suggests that his arrival in Aotearoa as a *Ptilinopus regina* might be indicative of Anthropocenic environmental change, as well as reflecting a means by which species might be able to protect biodiversity. As Davis and Watson explain, "For many of these [vagrant] species, the ability to disperse and increase their range away from warming, drying or otherwise increasingly unfavourable climates may buffer them against extinction if they can move fast enough" (239). Based on their assessment of the 2017 case of the Nicobar pigeon that landed in Australia, their stance is unambiguous: "The application of ad-hoc policies considering individual vagrants as a biosecurity risk is ill-informed, ecologically indefensible, and potentially counterproductive… It is, therefore, imperative that we allow species to disperse if we are to minimise global extinctions" (238–239). Davis and Watson also reflect on questions of scale, asserting that "[a]pplying lessons learned from invasive species management, we know that it is both the species AND the individual that matter" (240). Indeed, as Alexander Lees and James Gilroy have observed, while vagrant species are sometimes assumed to be "biological 'dead-ends' " (R1568), "a significant proportion may become the founders of new migratory routes and even populations" (R1570). Ori's story is therefore not just about the death of one bird, or a local extinction in New Zealand of an otherwise existing species; rather, he is a canary in a coalmine bringing a warning about how borders – many of which seem to be hardening in the face of the intensifying Anthropocene – are exacerbating planetary-scaled threats to human and non-human animals' entangled futures.

As the Anthropocene continues to make ever clearer species' reliance on one another, it is more crucial than ever that conversations are opened up that transcend taxonomic boundaries, and that extend empathy and respect to individual animal lives, instead of their just being rolled into a desubjectified inventory of types. As part of this, more dialogues will have to take place between the living and the dead – a practice that has long occurred in many different ways across a range of cultures and belief systems, but which Eurocentric museums have not historically made much space for. In thinking about Ori since I visited him, and in trying to put my responses to his story into words, I have been worried by questions of how he might also get to speak, or hold fast to his opacities in the silence that was imposed upon him. While I still remain at a loss with these questions, I continue to look for contexts in which related themes are negotiated. A particularly resonant body of work in this regard is multidisciplinary artist Helena Hunter's project *Falling Birds*, which crucially also indicates the potential of practices of mourning and memorialisation for reanimating extinction archives. In Hunter's project, which stemmed from a 2018 to 2019 residency at London's Horniman Museum, she spoke with, wrote with, and reimagined taxidermied bodies of extinct and critically endangered birds in the collection. Through having made x-ray images of them which illuminate their contrived internal frameworks and mounts (see Figure 2.5, for example), Hunter reveals the museological fictions and materialities that attempt to position casualties of environmental violence as display objects, and the project more broadly proposes speculative and personal modes of encounter that resist scopic orthodoxies and instead create room for the agency and subjectivity of the avian interlocutors. Through poems, Hunter examines the failures of language to gain purchase on the birds' experiences, and through patience, the projects' practices accept impasse – in contrast to perceived stasis – as part of the archival encounter.

Thinking about Ori in relation to *Falling Birds* is a reminder then that rites do not necessarily request answers, and that in contexts where mourning is to be a part of researching, specimens like OR.030538 might be memorialised not as an object of "discovery", but fundamentally – as Hannah Stark puts it – as "a creature whose life and death matters" (66). While I remain uneasy about putting myself into this discussion of Ori's story, this practice of bearing witness arguably necessitates it, and the uneasiness indeed partly stems from the persistence of the myth that it is possible to be an objective or disinterested observer within museological space. As I am reminded by the work of Hunter and others – indeed, by Ori – I was however already a part of this archival story, even before my visit to see him. If this thought is unsettling to me – as among other things a European immigrant living on indigenous land, and a person who has more progress to make in

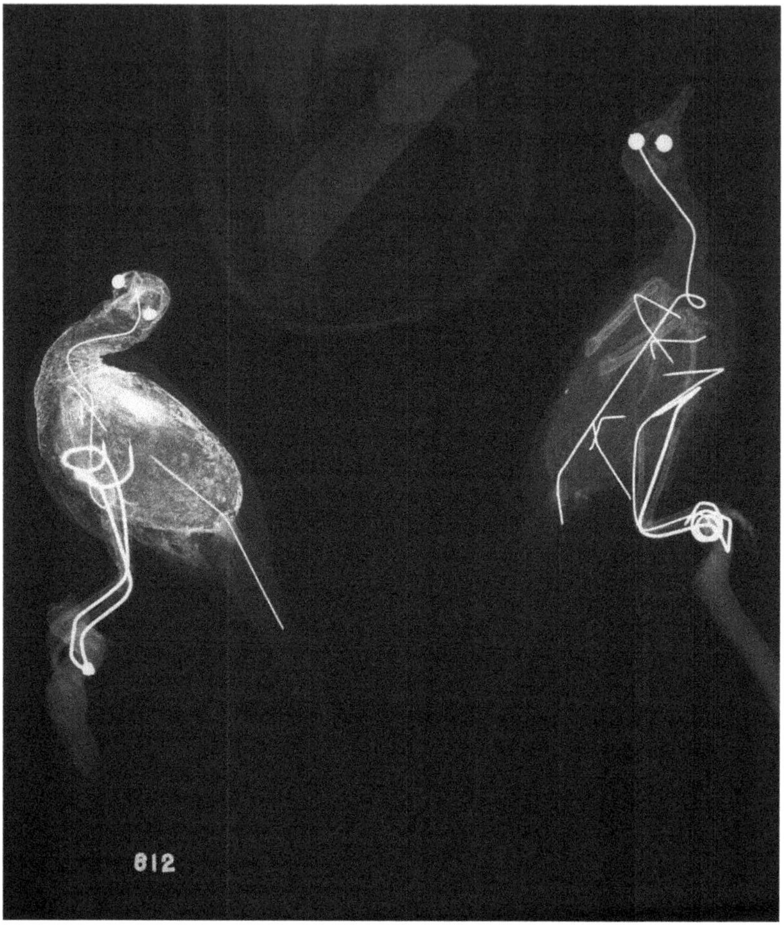

Figure 2.5 passenger pigeon x-ray image from *Falling Birds*, Helena Hunter, 2020. Image reproduced with the permission of the artist.

their efforts to rely less heavily on fossil fuels – then arguably it should be. In rupturing some of the narratives and illusions that can arise from museological contexts, perhaps Ori and other specimen animals might yet help to pierce the hermeneutical bubble of the archive, or at least enable some further circulation of insights and experiences in and out.

Notes

1 Te Whanganui-a-Tara refers to Wellington Harbour in te reo Māori, the indigenous language of Aotearoa New Zealand. However, this name is now also used to refer to the city of Wellington.

2 Presumably however this status as being of "least concern" could change rapidly – especially in light of the devastating forest fires that have swept vast tracts of Australia in recent years.
3 While the juvenile fruit dove's sex was unknown when I visited him, during a later stage of this research DNA-sexing undertaken by Te Papa revealed him to be male.
4 Indeed, as Sue Ann Prince has observed more generally,

> a specimen is a curious means of representing nature because what is used – whether a skeleton, a skin, or a dried flower – is composed of all or part of what was the living thing itself. It is thus more than a representation but less than real, live nature. It is mediated by human hands...
>
> (4)

5 Philip Armstrong has critiqued this pervasive and damaging notion in relation to sheep, and has demonstrated how their portrayal as numerous and fungible has led to their mistreatment by humans. When an individual sheep is celebrated, Armstrong observes, this is often because they are deemed to have acted in an atypical way; in departing from the rest of the flock, such an animal "became an individual and hence worthy of respect" (152).
6 While Thom van Dooren rightly cautions against the conceptualisation of extinctions as "singular events" (11), and observes that "the edge of extinction is more often a 'dull' one: a slow unravelling of intimately entangled ways of life that begins long before the death of the last individual and continues to ripple forward long afterward" (12), he also notes that there are "various 'edges of extinction'" for different species (12). Because Ori's death represented a local extinction of an otherwise extant species, it is possible – and sobering – to imagine that this loss might later be seen as somewhere within this "dull edge of extinction" (13) for the *Ptilinopus regina* as a whole. Indeed, arguably Ori's death – as something both abrupt and with slow-burning implications (illuminating, for example, the ways in which borders might stymie the establishment of new populations of species whose existing habitats are under threat) – indicates how specific occurrences shape extinction's "dull edge."
7 Indeed, in a 1990 article in the *Los Angeles Times*, Ron Askew suggests the role oil rigs have played in ornithology, with bird watching being a key form of recreation for workers aboard offshore platforms, and organisations such as the North Sea Bird Club gathering valuable data on bird distribution, including rare sightings such as the first black-billed cuckoo (typically resident in the Americas) to appear in Britain – a sighting that was accepted by the British Ornithological Union. While the Club's backing by various oil companies raises questions of corporate greenwashing, the rig workers' motivations for being involved in ornithology should not be assumed to be the same as those of the companies.
8 However, Jingyi Shen's research also suggests how camphor wood might be "a potentially harmful museum storage material" (46,458).
9 It is also interesting to note here that doves and mines are specifically linked in some cultural imaginaries. As observed in Barbara Allen's multifaceted study *Pigeon*,

> A commonly held superstition is that if a dove is seen near a mineshaft, there will be an accident in the mine... The appearance of a dove hovering near the mouth of a coal pit in 1902 [in Wales] was enough to cause 300 men to refuse to go to work there...
>
> (85)

10 Indeed, in a fascinating study published in 2017, Shane DuBay and Carl Fuldner illuminate the presence of atmospheric black carbon particles on bird specimens

in museum collections, and demonstrate how these skins can "provide durable snapshots of the past environments from which they were drawn" (11,325). By analysing the concentrations of this pollution on the bodies of 1,347 birds collected by museums over 135 years in the historically industrial states of Illinois, Indiana, Michigan, Ohio, Pennsylvania, and Wisconsin, DuBay and Fuldner were able to identify correlations between their findings and existing data on the shifting consumption of coal during this period of time – and the resulting emissions of black carbon. Furthermore, their research highlights the potential of bird specimens – or the environmental archives that inhabit them – to provide data that addresses gaps in historical records of atmospheric and environmental sampling. Summarising DuBay and Fuldner's work, JoAnna Klein observes that "Between 1880 and 2015, the filth on [birds'] feathers undulated with social changes and environmental policy. The dirtiest birds flew through the skies just before 1910, at the height of industrialization." A further reminder of the relationships between birds, air, and atmospheric composition is described by Allen:

> [i]n recent years pigeons have become smog monitors. In 2006 there was a plan to fit mobile backpacks to a flock of pigeons to monitor air pollution. Twenty pigeons would be released into the skies over San Jose, California, each carrying a GPS satellite tracking receiver, air pollution sensors and a basic mobile phone. Miniature cameras around their necks would also post aerial pictures. They would be messengers for the environment, continuing the role of their legendary ancestor, back in the time of the Flood. In that story, they were locating land; now they would be helping to preserve that heritage.
>
> (122)

Works Cited

Alberti, Samuel J. M. M. "Introduction: The dead ark." *The afterlives of animals: A museum menagerie*, edited by Samuel J. M. M. Alberti, University of Virginia Press, Charlottesville and London, 2011, pp. 1–16.

Allen, Barbara. *Pigeon*. Reaktion Books, London, 2009.

Armstrong, Philip. *Sheep*. Reaktion Books, London, 2016.

Askew, Ron. "North Sea oil rigs swarm with birds." *Los Angeles Times*, 19 Aug. 1990, www.latimes.com/archives/la-xpm-1990-08-19-mn-2702-story.html

"Bird airlifted to safety from North Sea rig released back into wild." *BBC News*, 1 Nov. 2013, www.bbc.com/news/uk-scotland-north-east-orkney-shetland-24771780

Callaghan, Corey T., et al. "Travelling birds generate eco-travellers: The economic potential of vagrant birdwatching." *Human Dimensions of Wildlife*, vol. 23, no. 1, 2018, pp. 71–82. https://doi.org/10.1080/10871209.2017.1392654

Carrington, Damian. "Arctic stronghold of world's seeds flooded after permafrost melts." *The Guardian*. 19 May 2017, www.theguardian.com/environment/2017/may/19/arctic-stronghold-of-worlds-seeds-flooded-after-permafrost-melts

Chrulew, Matthew. "Managing love and death at the zoo: The biopolitics of endangered species preservation." *Australian Humanities Review*, vol. 50, 2011, pp. 137–157.

Claramunt, Santiago, and Natalie A. Wright. "Using museum specimens to study flight and dispersal." *The extended specimen: Emerging frontiers in collections-based ornithological research*, edited by Michael S. Webster, CRC Press, 2018, pp. 127–141.

Crosby, Alfred W. *Ecological imperialism: The biological expansion of Europe, 900–1900*. Cambridge University Press, Cambridge, 1986.

Davis, Robert A., and David M. Watson. "Vagrants as vanguards of range shifts in a dynamic world." *Biological Conservation*, vol. 224, 2018, pp. 238–241. https://doi.org/10.1016/j.biocon.2018.06.006

Donnelly, Laura. "The owl who flew 2,000 miles – but finally arrived by helicopter." *The Telegraph*, 27 Nov. 2011, www.telegraph.co.uk/news/earth/wildlife/8917962/The-owl-who-flew-2000-miles-but-finally-arrived-by-helicopter.html

DuBay, Shane G., and Carl C. Fuldner. "Bird specimens track 135 years of atmospheric black carbon and environmental policy." *Proceedings of the National Academy of Sciences*, vol. 114, no. 43, 2017 pp. 11321–11326. https://doi.org/10.1073/pnas.1710239114

Evans, Mel. *Artwash: Big oil and the arts*. Pluto Press, London, 2015.

"Fact check: In the event of a fire, Yale library reduces oxygen levels in book stacks only." *Reuters*, 16 Feb. 2021, www.reuters.com/article/uk-factcheck-yale-library-fire-idUSKBN2AF1FN

Heatherington, Tracey. "From ecocide to genetic rescue: Can technoscience save the wild?" *The anthropology of extinction: Essays on culture and species death*, edited by Genese Marie Sodikoff, Indiana University Press, Bloomington, IN, 2012, pp. 39–66.

Hensley, Nathan K., and Philip Steer. "Introduction: Ecological formalism; or, love among the ruins." *Ecological form: System and aesthetics in the age of Empire*, edited by Nathan K. Hensley and Philip Steer, Fordham University Press, New York, 2019, pp. 1–17.

Hunter, Helena. *Falling Birds.* Horniman Museum and Gardens, London, 2020–2022.

IUCN. "Rose-crowned fruit-dove: *Ptilinopus regina.*" *IUCN Red List*, 1 Oct. 2016, Accessed 11 Nov. 2022, www.iucnredlist.org/species/22691430/93312183

Klein, JoAnna. "The dirty secrets saved in dead birds' feathers." *The New York Times*, 10 Oct. 2017, www.nytimes.com/2017/10/10/science/birds-air-pollution.html

Krebber, André and Mieke Roscher. "Introduction: Biographies, animals and individuality." *Animal biography: Re-framing animal lives*, edited by André Krebber and Mieke Roscher, Palgrave Macmillan, London, 2018, pp. 1–15.

Lees, Alexander C., and James J. Gilroy. "Bird migration: When vagrants become pioneers." *Current Biology*, vol. 31, no. 24, 2021, pp. R1568–R1570. https://doi.org/10.1016/j.cub.2021.10.058

"Lost racing pigeon Pedro gets helicopter ride home from oil rig." *Reuters*, 9 Jun. 2015, www.reuters.com/article/us-britain-pigeon-idCAKBN0OO1NL20150608

McKim, Claire. "Oil rig owl airlifted to safety by SSPCA." *The Scotsman*, 28 Oct. 2015, www.scotsman.com/regions/aberdeen-and-north-east/oil-rig-owl-airlifted-safety-sspca-1491295

Merriman, Nick. "Museum visiting as a cultural phenomenon." *The New Museology*, edited by Peter Vergo, Reaktion Books, London, 1989, pp. 149–171.

Miskelly, Colin M. "A new bird for New Zealand – rose-crowned fruit-dove." *Te Papa's Blog*, 8 Apr. 2020, blog.tepapa.govt.nz/2020/04/08/a-new-bird-for-new-zealand-rose-crowned-fruit-dove/

Miskelly, Colin M. "First record of rose-crowned fruit-dove (*Ptilinopus regina*) from NewZealand." *Notornis*, vol. 67, 2020, pp. 564–567.

Miskelly, Colin M., et al. "Vagrant and extra-limital bird records accepted by the Birds New Zealand Records Appraisal Committee 2019–2020." *Notornis*, vol. 68, 2021, pp. 253–265.

Morton, Timothy. *The ecological thought*. Harvard University Press, Cambridge, MA and London, 2010.

Mosher, Dave. "Oil rigs may turn migratory birds into shark food." *Wired*, 10 Jan. 2012, www.wired.com/2012/01/birds-sharks-oil-platforms/

"North Sea oil platform owls released back into the wild." *BBC News*, 20 Nov. 2015, www.bbc.com/news/uk-scotland-north-east-orkney-shetland-34880521

"Off-course migratory bird rescued from oil rig." *The Scotsman*, 1 Nov. 2013, www.scotsman.com/news/course-migratory-bird-rescued-oil-rig-1554595

"Osprey hurt in storms nursed on North Sea vessel." *BBC News*, 29 Sep. 2012, www.bbc.com/news/uk-scotland-north-east-orkney-shetland-19759847

"Otago Institute." *Otago Witness*, 24 May 1894, p. 21.

Ozone Secretariat. *Handbook for the Montreal Protocol on substances that deplete the ozone layer (14th edition)*. UN Environment Programme, 2020.

Poliquin, Rachel. *The breathless zoo: Taxidermy and the cultures of longing*. The Pennsylvania State University Press, University Park, PA, 2012.

Preziosi, Donald. "Brain of the Earth's body: Museums and the framing of modernity." *The rhetoric of the frame: Essays on the boundaries of the artwork*, edited by Paul Duro, Cambridge University Press, New York, 1996, pp. 96–110.

Prince, Sue Ann. "Stuffing birds, pressing plants, shaping knowledge: Natural history in North America, 1730–1860." *Transactions of the American Philosophical Society*, vol. 93, no. 4, 2003, pp. 1–9.

Ronconi, Robert A., et al. "Bird interactions with offshore oil and gas platforms: Review of impacts and monitoring techniques." *Journal of Environmental Management*, vol. 147, 2015, pp. 34–45. https://doi.org/10.1016/j.jenvman.2014.07.031

Scott, Heidi C. M. "Industrial souls: Climate change, immorality, and Victorian anticipations of the good Anthropocene." *Victorian Studies*, vol. 60, no. 4, 2018, pp. 588–610. https://doi.org/10.2979/victorianstudies.60.4.04

Shen, Jingyi. "Camphor wood, a potentially harmful museum storage material: An analytical study using instrumental methods." *Environmental Science and Pollution Research*, vol. 28, 2021, pp. 46,458–46,468. https://doi.org/10.1007/s11356-020-09446-0

Stark, Hannah. "The cultural politics of mourning in the era of mass extinction: Thylacine specimen P762." *Australian Humanities Review*, vol. 63, 2018, pp. 65–79.

Stuchel, Dani. "Material provocations in the archive." *Journal of Critical Library and Information Studies*, vol. 3, no. 1, 2020, pp. 1–25. https://doi.org/10.24242/jclis.v3i1.103

Te Aka Māori Dictionary, Accessed 31 Oct. 2022. www.maoridictionary.co.nz

Tuysuz, Gul, and Arwa Damon. "Arctic 'Doomsday Vault' opens to retrieve vital seeds for Syria." *CNN*, 19 Oct. 2015, www.edition.cnn.com/2015/10/19/europe/svalbard-global-seed-vault-syria/index.html

van Dooren, Thom. *Flight ways: Life and loss at the edge of extinction*. Columbia University Press, New York, 2016.

Wagner, Bayliss. "Fact check: Yale library's fire system protects rare books without suffocating people." *USA Today*, 12 Feb. 2021. www.usatoday.com/story/news/factcheck/2021/02/12/fact-check-yale-beinecke-library-fire-suppression/4461144001/

Webster, Michael S. "The extended specimen." *The extended specimen: Emerging frontiers in collections-based ornithological research*, edited by Michael S. Webster, CRC Press, Boca Raton, FL, 2018, pp. 1–9.

"Winsford Salt Mine: Plan to extend hazardous waste storage until 2045." *BBC News*, 4 May 2022, www.bbc.com/news/uk-england-merseyside-61317857

3 Listening to Lost Species

Memorialising Extinction through Sound

Hannah Hunter

What does extinction sound like?

We might imagine a dinosaur-decimating asteroid as a quick burst of incredible, unfathomable noise, followed by a long period of excruciating silence. Before this ear-splitting, epoch-ending impact, we might speculate upon the bellows, the cries, the thumps, and the scuttles of those prehistoric creatures that we now know only through traces and fictions. For more recent extinct species, we may have a better idea of their sounds: a passenger pigeon, for instance, likely sounded at least similar to our contemporary cooing neighborhood residents, though written accounts also describe "shrieks and chatters and clucks" (Craig 408), that, when heard *en masse* from the species' characteristic giant flocks, could be "deafening" (Fuller 23). For some recently extinct species, we even have recordings of their voices: since the advent of portable sound recording in the early twentieth century, some creatures have had the dubious honor of having their sounds immortalized by humans before their human-induced demises. For instance, the Baiji (Yangtze River Dolphin) species has not been credibly sighted since 2004, but the posthumous whistles of Qi Qi—a Baiji who was rescued into captivity in 1980 after becoming injured by fishing equipment—are available for listening online (*Baiji, Yangtze River Dolphin*). Recordings like these offer a novel and powerful connection with lost animals—never before have we been able to hear directly from extinct beings themselves. But what might they have to tell us, and how might we be urged to respond?

Listening to sound recordings of lost beings has the potential to produce powerful empathy. As noted by Derrida in a discussion about sound recordings of deceased humans:

> I am always overwhelmed when I hear the voice of someone who is dead, as I am not when I see an image or a photograph of the dead person… I can, *here and now*, be affected by a voice from beyond the grave.
>
> (70)

Here, Derrida is clear that this affective potency of sound recordings hinges on their collapsing of temporalities: on pressing play, voices from that most

DOI: 10.4324/9781003315957-4

distant of geographies—beyond the grave—are conjured in the present time and place. And yet, the ephemerality of sound also has another effect: it is always already slipping away. Sound recordings of past events, people, and, as this chapter explores, animals are ghostly presences in our world, allowing for intimate but evanescent relations with that which is no longer. In this opening, however brief, we might find creative and political potential for mourning and even resisting mass extinction. Sound both amplifies and disavows loss—it connects, conjures, and tears apart.

These affective possibilities of extinct voices have been taken up by several artists and curators grappling with our ecological crisis. For instance, artists have created musical compositions featuring extinct voices and created sculptural speculations on those voices that were never recorded. Many of these efforts deal with species who became prematurely extinct due to human action, and some artists are explicit that they hope to not only create a space for mourning but also inspire listeners to take action to prevent future extinctions. Together with the growing archive of extinct animal sound recordings, these works and practices are efforts in *giving voice* to lost beings as part of a broader genre of *sonic memorials to extinction*. This genre has a significant impact on how extinct animals are remembered and how extinction is narrated in public spaces. As such, it demands critical inquiry.

In this chapter, I critically consider the role of sound, particularly audio recordings and sound art, as a medium for communicating, mourning, and resisting animal extinction. I first explore the concept of sonic remembrance in more detail, something that has become more popularized in Western memorial culture, including in ecologically orientated memorials, in recent years. I argue that the connective possibilities of sound recordings and sound art, as well as their relative novelty in memorial spaces, lend them as particularly effective mediums for memorialization. Focusing specifically on works and practices in the United States (with one example from a French artist) that attempt to "give voice" to extinct species, the remainder of this chapter then explores the promises, potentials, and politics of sonic memorials through three overlapping categories: (1) recording vanishing voices; (2) creating with extinct voices; and (3) speculating upon extinct voices.

I consider case studies that range from a 1935 bird sound recording expedition that sought to capture America's disappearing species, to a recent series of physical "sound sculptures" representing extinct and endangered animal voices, to a web installation that speculates upon the sonic worlds of lost species. I argue that, beyond their common subjects, these works and practices are united in their efforts to preserve, evoke, and create sonic connections between humans and extinct animals. In so doing, sonic memorials to extinction do not only mourn that which has been lost, but create spaces of ongoing relation, speculation, refusal, and, most of all, reflection. What have we done? What are we to do? What could/ should we do?

In a period that is characterized by "a rapidly increasing series of extinction and climate change events on the one hand, and an equally rapidly accelerating scientific and technological capacity to reengineer life and living materials, on the other" (Stephens 872), the temporalities of extinct voices are knottier than ever. I argue that it is this meeting of the time-bending affordances of sound with the complexities of this period where sonic memorials have the potential to transform our relations with lost species. At the same time, however, I am critical of approaches that blindly celebrate the connective possibilities of sound without considering, for instance, the wider contexts of wildlife recording practices.

Sonic Remembrance

When one thinks of Western memorials and monuments at large, these usually cater to audiences' visual sensibilities, for instance, a statue of a deceased figure, a commemorative plaque for a past event, a grand building named after a notable deceased patron. However, though far from mainstream,[1] sound is being increasingly utilized in public memorial culture. In the aftermath of 9/11, for instance, the Sonic Memorial Project circulated widely: a digital sound archive containing "stories, ambient sounds, voicemails, and archival recordings to tell the rich history of the twin towers, the neighborhood, and the events of 9/11" (*Sonic Memorial*). Additionally, nearly twenty years after the attacks, the Tower of Voices was erected at the Flight 93 National Memorial in Pennsylvania, a "ninety-three feet tall musical instrument holding forty wind chimes, representing the forty passengers and crew members [aboard the flight]" (*Tower*). Sound has also featured in Holocaust memorials, such as the University of Southern California Shoah Foundation's *Dimensions in Testimony* website, where users can engage in a "virtual conversation" with Holocaust survivors. Such physical and digital memorials demonstrate the power of sound as a medium to remember, evoke, and mourn past events and people.

In recent years, sound has also featured in public works and practices addressing mass extinction. For instance, American artist Maya Lin's "last memorial" series, *What is Missing* (2009–), employs both visual and sonic registers to evoke what we have lost, and what we are losing, in the extinction crisis (DeLaure). In Lin's work and others', sound recordings of endangered and extinct species are employed as haunting specters of humanity's past wrongs and urgent calls to action (DeLaure). Other works like Juan Oliver and Crystelle Vu's *Extinction Gong* (2017) use sound as less a representation of particular extinct or near-extinct creatures, but as a marker of incessant death: a mechanism automatically beats the gong every nineteen minutes, the approximate "rhythm of species extinction" (Oliver & Vu). "Should biologists declare a new species extinct while the Extinction Gong is active" the artists explain, "it will receive an update...and perform a special ceremony: four strikes in quick succession alongside a text-to-speech utterance of the Latin name of the species lost, resonating through the gong" (Oliver & Vu). These sonic apparitions cut through traditionally visual spaces like galleries and museums as if to wake audiences up from our apathetic slumbers: they call out and, ideally, we are urged to respond.

Indeed, sound has proven an affective force for communicating our ecological crisis ever since Rachel Carson's *Silent Spring* rallied American publics to action around the imaginary that unfettered ecological abuse could cause a spring without birdsong. Around the same time, Roger Payne and Scott McVay's album of recorded whale music, *Songs of the Humpback Whale*, became popular with middle-class urbanites as a form of sonic self-care and an evocative tool for the environmental movement (Ritts). More recently, sound art theorist and composer Jonathan Gilmurray has observed the emergence of the genre of Ecological Sound Art—works that

> [use] sound as a medium not just to raise awareness of ecological issues, but to help us to understand them from new perspectives, relate to them in new ways, and reconnect with them in ways that might just motivate us to act.
>
> (Gilmurray "Sounding" 72)

For instance, in protest of proposals to build a major hydroelectric power plant on the Baker River in Chile, Chilean artist Graciela Muñoz' installation *El Sonido Recobrado* played soundscape recordings of the Baker on the floor of a different Chilean river that had been dried up after illegal industrial draining and damming (Gilmurray *Ecology*). Other examples of this genre include Andrea Polli's *Sonic Antarctica*, an album weaving Antarctic field recordings with audio interviews with concerned climate scientists (Gilmurray *Ecology*), and Rachel Belmore's 2017 *Wave Sound,* a series of "listening cone" sculptures placed in natural areas in Canada that invite audiences to listen to the land and consider their relationship with it (Belmore in Nixon).

In Gilmurray's writing and in others', sound and sound art are considered powerful, but under-theorized, tools for communicating and sensing the troubles of the Anthropocene (Barclay; Comstock & Hocks; Kanngieser; Whitehouse). Though many are careful not to over-emphasize its sensorial distinctiveness, sound studies scholars have described the intensive, immersive, and affective qualities of sound as a vibrational force that can literally move bodies, transforming relations between disparate individuals—human and otherwise (e.g. Born; Browning; M. Gallagher "Field" and "Sound").[2] As such, sound might be particularly useful in raising awareness and promoting action for the environmental crisis, given its ability to make connections across bodies, species, spaces, and temporalities. Though we are routinely flooded with visual representations of ecological devastation—starving polar bears, out-of-control wildfires, giant ocean garbage patches—sound can be as, if not more, striking in communicating ecological damage (Comstock & Hocks). Regardless of the extent to which listening to the ecological crisis may be more or less effective than looking at it, it is at least true that sonic forms of communication offer novelty and alterity at a time where audiences may find themselves oversaturated and exhausted by the repetitive visual canon of climate change. Sound can thus bring fresh attention and urgency to environmental problems, connecting us to disparate species, spaces, and disasters in distinct ways.

I take the diffuse acts and works discussed in this chapter as evidence of a burgeoning genre of *sonic memorials to extinction*, something that has emerged in part because of the relatively recent explosion of wildlife sound recording and circulation, as well as a widespread push to enliven museum and gallery species with immersive sensory experiences. Though there are many works that might comprise this genre—including those that mark the crisis like *Extinction Gong,* and those that lament it through music like Sarah Kirkland Snider's *Mass for the Endangered*—in this chapter, I focus on works and practices that attempt to *give voice to* extinct species. These offer perhaps the most explicit attempts to memorialize animal extinction as an issue entangled but distinct from other ecological devastations, and are also an evocative point of comparison for other kinds of species-specific extinction memorials, like thematic taxidermy displays (see, e.g., Bezan; Enright). Moreover, historical legacies of collection and display, emerging forms of technology, and technocultural desires for de-extinction are propelling these particular kinds of sonic extinction memorials in intriguing ways.

Recording Vanishing Voices

The first mechanical sound recording of wild birds in North America was taken by Cornell University ornithologists Arthur A. Allen and Peter Paul Kellogg in 1929: a short burst of a Song Sparrow melody, captured in Ithaca's Stewart Park (T. Gallagher "Birth"). Allen, the founder of the Cornell Lab of Ornithology (CLO), had the anxiety of bird loss in his heart from the very beginning. Realizing the memorial potential of this new technology, he quickly organized a cross-country bird sound recording and photography expedition with the American Museum of Natural History (AMNH), the motivations of which are worth quoting at length:

> Years have passed since the cooing of the Passenger Pigeons was a common sound throughout eastern North America. The tooting of the Heath Hen, the quack of the Labrador Duck, and the grunt of the Great Auk are sounds which have passed forever from this good world and even from the memory of man. Still other sounds of nature which were familiar to our ancestors are now seldom, if ever, heard, and many of these promise soon to be entirely lost.
>
> It was with this thought in mind that the members of the staff of the Laboratory of Ornithology at Cornell University… decided to make an expedition in the spring of 1935, the object of which should be to perpetuate for all time the voices of such American birds which, because

of their rarity, have given promise of soon becoming but forgotten memories.

(Brand et al.)

Here, Allen and his colleagues are explicit in their intentions to sonically memorialize birds: by recording the sounds of disappearing species, they hoped to guard future generations against the pain of losing not only the nation's birds, but also their voices.

Alongside collecting the sounds of endangered birds that persist today—like snowy egrets and golden eagles—the crown jewel of the 1935 expedition was the first ever sound recordings of the iconic Ivory-billed Woodpecker (the "ivorybill"). Until this small population of ivorybills was found in Louisiana a few years prior to these recordings, the species was widely presumed extinct in the United States, and its absence/presence has been the subject of intense controversy ever since. Known as the "Grail Bird" or "Lord God Bird", the ivorybill is uniquely revered in the ornithological community, with many birders and scientists devoting significant time and resources to try to rediscover the species (T. Gallagher, *Grail*). Indeed, despite several alleged captures since then, the 1935 recordings remain the only undisputed sound recordings of this species. A recent US Fish and Wildlife Service proposal to officially recognize the ivorybill as extinct and thus remove it from the Endangered Species Act List (USFWS) seems to calcify what many already felt in their guts: the crackly, nasal ivorybill calls recorded in 1935 are some of the only traces that remain of the Lord God Bird.

Of the eleven bird species the USFWS recommended to be removed from the Endangered Species Act List in 2021, there are known sound recordings for six of them. The ghostly voices of not only the ivorybill but also the Bachman's warbler, Bridled White-eye, Kaua'i 'ō'ō, Kāma'o, and Po'ouli are freely available for listening on CLO's *Macaulay Library* website (Hunter). The cultural power of these recordings is particularly apparent in the case of the Kaua'i 'ō'ō, a small songbird endemic to the Hawaiian island of Kauai'i. There are several recordings of this species' intricate song on the *Macaulay Library* website, but a particularly famous one was captured by ornithologist Jim Jacobi in 1986. This recording depicts a lone male Kauai'i 'ō'ō, a dueting species, singing out to a mate, with pauses in between meant for a female to reply. Instead, there is silence: seemingly, no one is left to respond. Recordings of lone males have circulated widely online, with one YouTube remix of a Kaua'i 'ō'ō recording accumulating almost 2 million views at the time of writing this chapter ("Kauai 'O'o"). Comments left underneath the video are unusually ruminative, with listeners lamenting the loss of this species and the sorrow they feel can be heard in the lone male's voice (Steensen). Indeed, with my own interviews with scientists and archivists about extinct animal sounds, several brought up their strong emotions towards the Kaua'i 'ō'ō recordings. One ornithologist even told me that he struggled to listen to them without crying. Such responses to the Kaua'i 'ō'ō's haunting song demonstrate the power of sound in communicating the tragic losses of species extinction.

Though the 1935 Cornell-AMNH expedition is well known, the archival papers of CLO associates reveal that capturing the voices of diminishing species was a major motivation for many early sound recordists. For instance, Arthur A. and Ellen Allen devoted significant effort to capturing the voices of Whooping Cranes in Arkansas in the 1950s (Kellogg), and in 1936, Kellogg expressed a desire to attempt to photograph and possibly record the Carolina Parakeet—a species that, his correspondent reminded him, "has been extinct some 15–30 years!" (Brand). Though, as in 1935, such impulses to record rare species were partially archival ones, for this community of recordists sounds were never collected *only* as ecological relics for future human ears. Rather, these recordings were captured with either immediate or speculative conservation purposes.[3]

For instance, in 1961, CLO associate George B. Reynard recorded the Puerto Rican Whip-poor-will (also known as the Puerto Rican nightjar), a species that ornithologists in the Continental United States had long presumed extinct. Subsequent searches in the recording area found that a remnant population of this species had survived (Reynard). In these searches, audio recordings were not only a catalyst but a major search tool—Reynard played back the recordings to "the oldest people living in the area" (Reynard 51) to survey if they had heard the species before, and broadcasted the recordings in potential habitats to lure remnant nightjars into nets. The 1935 ivorybill recordings have similarly been used in "playback" techniques to attempt to relocate the species (T. Gallagher *Grail*).

Clearly, sound recordings are quite different kinds of traces of extinct beings than, say, a taxidermy mount or a photograph. Though all kinds of traces can and have been used to aid conservation or rediscovery efforts,[4] sound recordings are unique in their time-bending and necessarily performative capacities: playing a historical sound recording may invoke a historical sonic moment, but also always enacts a new one (M. Gallagher "Field"). This means that a historical ivorybill recording can be replayed into a modern forest *as if a modern ivorybill was really calling*, with the potential to trick living birds into believing a living ivorybill is actually there. When humans hear recordings of extinct birds, a similar form of mental trickery takes place—we know they are gone, but their voices tumble into present times and spaces. Like in Derrida's reflections, we can be affected by them "*here and now*" (Derrida, 70). In this zone of sonic intimacy, as demonstrated in the responses to the Kauai o'o recordings, we might find a deeper empathy with these lost creatures, who become temporarily enlivened and individualized through sound. At its strongest, this empathy could rouse listeners to action. Indeed, cogitative research has found that feelings of empathy towards animals can, under certain conditions, be an important motivator for environmentally positive behavioral change (Young et al.).

Though there is much to celebrate about the connective and affective possibilities of sound recordings of lost animals, it is important to note that they are not neutral captures. Like taxidermy mounts, paintings, and photographs (Haraway; Ibbotson; Ryan), sound recordings are curated, framed, and edited socionatural objects, co-created by the animal, their environment, the recording technologies, and the positionality, subjectivity, and skills of the recordist. For instance, the ivorybill recordings were captured with the aid of a parabolic reflector, a large satellite-dish-shaped object used to concentrate and amplify the sound source the reflector is orientated towards. The result is a "clean", focused recording with minimal background noise. Such practices, sound studies scholar Joeri Bruyninckx argues, are evidence of recordists' efforts to "sanitize" sounds and make them fit the scientists' laboratory sensibilities.

The positionality of these twentieth-century recordists (often white, male, settler American scientists) makes a difference here too—not only affecting the framing of the recording, but also how we might listen to it, knowing that the recordings were taken from the unceded lands of various Indigenous groups. Indeed, many US sound recordings of extinct animals come from Hawaii, a place subject to devastating ecological losses ever since European colonization (Boyer). The extinction of the Kauaʻi ʻōʻō, for instance, is attributed largely to introduced predators and colonial agricultural practices (Steutermann Rogers). Especially given the cultural and monetary value of rare sound recordings, it is crucial to consider Indigenous data sovereignty in relation to acoustic data extraction in such places, something by no means limited only to historical activities (Kukutai & Taylor; Ritts & Bakker 152; Robinson).

These points are important to keep in mind regarding the sound artworks that will be discussed in the rest of this chapter, some of which directly employ these audio recordings towards presumed progressive ends. This is not to say that the conditions of these recordings' capture diminish their political capacities, but rather that their entanglement in systems of oppression and extraction, as well as with Western scientific systems of framing and ordering, must not be ignored.

Creating with Extinct Voices

Aside from the bustling of visitors and the occasional audiovisual insertion, museums and galleries are usually quiet places. However, there is an increasing push to make these spaces more immersive and experimental, with many exhibits incorporating multimedia elements, including sound (Eley; Cortez). In addition, the large-scale digitization of the *Macaulay Library* and other sound collections since 2000 has allowed more people to access and be inspired by animal sound recordings in their works. In this section, I will discuss two recent works that utilize the sounds of extinct and endangered animals, *Requiem* (2018) and *Tipping Point: Echoes of Extinction*

(2020), asking how, and with what effects, they memorialize and give voice to lost beings.

Requiem is an audio installation by American sound artist and musician Steve Norton. The work comprises several speakers placed in a circle that play a composition of twelve sound recordings of recently extinct birds and frogs sourced from the *Macaulay Library* and *The Amphibian Foundation*, including the Bachman's Warbler, the Golden Toad, and the Kāmaʻo. An accompanying booklet lists each species along with their range, date of extinction, causes of extinction, and information about who recorded each sound and where. The installation is meant for "an empty and preferably darkened space", which Norton hopes will "focus attention on the sounds of the piece" (Norton). All of the species heard, as with so many recent extinctions, were prematurely extinguished due to human actions.[5]

The title of *Requiem* is evocative, gesturing to religious and musical practices of remembering the dead and, in the Roman Catholic church at least, reposing their souls. Requiems feature strongly in musical history: most famously, Mozart's haunting *Requiem in D Minor k. 626* (1791) was composed to commemorate an aristocrat's deceased wife. By reclaiming this form for extinct animals, Norton undermines human exceptionalism by making a clear declaration about these species' grievability and creating sonic space for this grief to be felt. Indeed, scholars have recently taken up Judith Butler's concept of "grievable lives" (Butler) in relation to more-than-humans, arguing that recognizing and making space for grief about animal death is politically productive (Stanescu). There are several pieces that employ similar symbolism in relation to nonhuman extinction, for example Snider's *Mass for the Endangered* and Enjott Schneider's chamber music composition *REQUIEM. About Insects & The Microbiotic World.*

What sets Norton's piece apart, however, is the use of the animals' *own voices* as the medium of our lamentation. This piece thus both *declares* and *allows for* these creatures' grievability: listening to lost voices in a darkened room demands a kind of affective connection with these beings that is difficult to ignore and, at times, difficult to endure. In fact, Norton claims that "people generally find it beautiful and relaxing. But as they think a little harder about what's going on, they often get very sad. Sometimes there are tears" (in Marshall). His hope is that audiences will "take [those emotions] with them and try to figure out what it is that they can do to contribute to the solution of these issues" (ibid.). In this way, *Requiem* uses sound art as a medium both for remembering past extinctions and for inspiring action to halt future ones.

American artist and sculptor Elizabeth Turk's *Tipping Point: Echoes of Extinction* sonically memorializes extinction quite differently, through physical "sound columns". These are 3D sculptural representations of extinct and endangered animal sound waveforms— a common (usually two dimensional) form of visualization for acoustic signals. As if in response to critical scholars who discuss the role of visual representations in flattening

Figure.3.1 "Sound columns" from Elizabeth Turk's *Echoes of Extinction* series. Photo credit and permission: Elizabeth Turk.

and objectifying sound (e.g. Mundy; Bruyninckx), Turk's sculptures transform these data into something evocative, corporeal, and, ultimately, beautiful. For instance, Turk's *Ivory-billed Woodpecker* (2020) sculpture is crafted from walnut and is 72.5 inches (approximately six feet) high and sixteen inches in diameter. Standing vertically, as opposed to the usual horizontal orientation of a waveform, the totem-like sculpture consists of a wooden pole with discs that periodically increase and decrease in size to the rhythm of the ivorybills' distinctive nasal calls: "kent… kent… kent… kent… kent kent… kent". In contrast, the *Gould's Emerald* sculpture is created from black anodized aluminum, which, molded into the more complex sounds of this extinct songbird, creates a fuzzy, uncanny effect (Figure 3.1).[6]

Even without the QR code links to sound recordings that accompany these sculptures in exhibition, the sonic effects of the sculptures are palpable: the physical forms take up some of the space once occupied by the bird calls themselves. Bringing presence to absence has been a core component of much of Turk's artistic practice, who considers her works with marble to be "a reductive process producing rich, negative space… emphasis[ing] what is not there, that which no longer exists" (Turk). Turk hopes that *Tipping Point* will start conversations around the question "Are we creating a silence?" Rather than only offering an opportunity to mourn, the title of the exhibition offers the potential for hope, as "tipping points have more than one direction" (Turk). Indeed, the slim, vertical sculptures seem to mirror the precarity of the animals they represent: some are suspended from the ceiling, and some are connected to the floor at their thinnest (quietest) points. For some species in the exhibition, their fate seems to be tilting in a positive direction, for instance, the Bald Eagle, which is included as "an optimistic story" (Turk in Greenberg). Still, anxiety and devastation can't help but hang heavy over these sculptures: a year before they were first exhibited in the Hirschl & Adler Modern, an article in *Science* estimated that the population of US and Canadian birds had declined by almost 3 billion birds since 1970 (Rosenberg et al.). Recent studies have shown aquatic species, who also feature in *Tipping Point,* as perhaps even more vulnerable, with many species becoming extinct due to human-induced environmental changes before they have even been discovered (Bastian; Pinsky et al.). These voices, and many more, have been or will be lost.

We might think about works that broadcast extinct creatures' voices with the concept of *invocation.* To invoke is to conjure: to call out for someone/something, to invite, to bring it to presence. In a chapter about the *Say Her Name* movement, Imani Danielle Mosley explains how musicians and protesters voice the names of Black women victims of police brutality. Mosley argues this "invokes and gives life to previously unheard voices" in order to raise awareness and inspire action (Mosley 144). Here, voicing the names of murdered and assaulted Black women is an act of active remembrance insofar as it is an act of refusal: refusing to remain silent, refusing to forget, refusing to abstract, and refusing to let state violence against Black women continue. The sonic invocation of works like *Requiem* and *Tipping Point* is of course quite different—addressing different issues to different ends—but shares a similar intention to *take up sonic space* by means of *giving voice to* that which is lost or hidden. In the art memorials discussed in this section, extinct animals' voices take up space in distinct ways, composing the presence of their absence to draw attention to the absence of their presence. The presence of their voices—either in sound recordings or in sound sculptures—is necessarily uncanny, allowing audiences a closeness with each being while also ripping them away. In this sonic space, audiences are confronted not with the abstract data of climate change but with the uncanny invocations of its individual ghosts—a loss that we can feel "in our gut" (McKibbon in Gilmurray *Ecology* 19).

In this space of uncomfortable intimacy, audiences would ideally be moved to take action to avoid future extinctions, as both *Requiem* and *Tipping Point* intend to promote. However, the impact of these works on audiences' behaviors is difficult to measure. Who gets to experience these artworks, and how are their wider lives entangled with extinction-accelerating processes? What sorts of consequent actions would be appropriate and meaningful? Such questions are crucial for both artists and audiences to grapple with. Part of the power of these works, though, is that they provide space for such questions to be asked: they refuse the silencing of extinct species, and, in doing so, create spaces of ongoing encounter, relation, and reflection.

Speculating upon Extinct Voices

But what of those species that expired too soon, whose voices were lost to memory? Many of those animals that feature most prominently in extinction narratives never had their voices mechanically recorded: the sounds of Dodos, Woolly Mammoths, and Passenger Pigeons are merely objects of informed speculation. And many *do* speculate, actively, upon these lost voices: artists and scientists have devoted significant effort to imagining and reconstructing the voices of these species, hoping to enliven our knowledge of lost beings through their sounds. In this section, I will consider two sound artworks that speculate on extinct voices: *Echozoo* (2020) and *Opera of Prehistoric Creatures* (2012).

Echozoo is a digital sound installation by American composer and sound artist Michael Reiley (*Micronesia*). On the website's landing page, users can select from six continents on a map and enter a speculative sonic world of extinct creatures and their habitats. For instance, selecting South America will lead you to the imagined sounds of seven creatures including the armadillo-like Glyptodon, extinct since ~4000 BCE, and the Candango Mouse, extinct since ~1960 CE. In the Asian continent, Reiley's composition for the Steller's sea cow, a large aquatic mammal found in the North Pacific until 1768, is an eerie immersion into the underwater world: the low grunts and bellows of the sea cows resonate through the thick flows of the ocean. As listeners, we are submerged alongside these strange aquatic mammals, bubbles, and ripples transporting us into deep waters from long ago.

All of the compositions that comprise *Echozoo* were modeled from the species' extant relatives, recorded by Reiley himself. However, rather than striving for scientific realism, "[t]hese long-silenced voices are re-sounded and heard within an imagined soundscape of their habitats" thus offering "an invitation to listen from the perspective OF the extinct creature" (Reiley). The result is "an acoustic 'augmented reality'" that intends to connect listeners with past creatures through both the animals' *voices* and their *ears*, where "listening [is] an experience of deep and profound empathy" (Reiley). Not unlike the works considered in the previous section, Reiley is explicit that *Echozoo* is not only "a memorial both to these creatures, and the earth as we have known it until now", but also a call for improved human-nature relationships in the present. Speculatively listening to and through extinct animals is thus another practice in making them present, making them grievable, and affecting change. By sonically transporting us into the extinct animals' *own sound worlds*, however, Reiley does something distinct in attempting to reach across not only temporal distances but also the human/animal divide that characterizes the postindustrial Global North—a violent division many argue to be at the root of our current ecological crisis. Speculatively listening *as* these creatures thus disrupts the illusion of human exceptionalism, offering the potential to not only afford a deeper, embodied empathy with the work's extinct subjects, but also with nonhumans more broadly.

French artist Marguerite Humeau's sculptural-sonic work *The Opera of Prehistoric Creatures* takes the challenge of "reconstructing" extinct animal voices more literally. In this work, Humeau created hypothetical sonic reconstructions of three prehistoric creatures: an Ambulocetus (*Walking Whale*), an Entelodont (*Terminator Pig*), and a Mammoth Imperator (Humeau in Debatty). To bring these creatures' voices to life, Humeau created sculptures of the vocal apparatus of each species: their "larynx, vocal chords, trachea, lungs, resonance cavities" (Humeau in Debatty). Air is then forced through the sculptures, vibrating through the moulds to create eerie groans, hums, and squeals. The challenge of creating this work, however, was that soft tissue does not fossilize, so the exact composition of these animals' vocal assemblages is unknown. As such, the sculptures were informed by "different areas of science, but also, and most important, other tools like speculation, design, rumours, collective imaginary" (Humeau in Debatty).

Humeau describes the resulting sculptures as "semi-real, synthetic ruins" (ibid.). Like in *Echozoo,* absolute realism was not the goal here, something made obvious by the visual construction of the sculptures, which are clearly not attempts at anatomical accuracy. Rather, the "disembodied vocal tracts" and abstracted skulls presented are "ambiguous" (Stephens 872). These skeletal, precarious objects are reminders of how much is unknown about the beings they represent, and how much we have lost—in fact, Humeau deliberately selected creatures that were less "known" to Western science (Humeau in Debatty). As cultural theorist Elizabeth Stephens notes, works like Humeau's are "provocations rather than explanations; they are intended to pose questions, rather than provide answers" (Stephens 882). Stephens considers such provocations in the context of de-extinction, the (still largely speculative) scientific efforts to "bring back" extinct species,[7] arguing that art-science collaborations like *Opera* are exercises in "speculative biology" (Stephens). As opposed to the extinct/extant, life/death binaries that inform mainstream de-extinction narratives, artworks like Humeau's "attempt to reimagine and redesign the biological in response to its current state of precarity" (Stephens 871).

From this perspective, *Opera* is less a traditional memorial for what is lost, than a provocative and playful experiment with what remains, and what might be to come. Indeed, the temporal distance between us and the species represented demands different responses than those in *Echozoo* or *Requiem*: feeling grief and anguish for the loss of Ambulocetus, creatures that lived around 48 million years ago, does not seem appropriate or productive. Importantly, all the other works discussed in this chapter have dealt with species lost, in part, *because of human action*. Part of the intended impact of these works is an internalization of responsibility and a push for action. It is difficult to know how to feel, then, when listening to the speculative voices of species that went extinct well before the evolution of humans. Sorrow? Wonder? Unsettlement? Hope? Particularly when encountered in the shadow of our current extinction crisis and shaky promises of de-extinction, *The Opera of Prehistoric Creatures* brings these creatures closer to us than we are used to, closer than we might be prepared for. Their unsettling, provocative form gestures not only to the promises but the limitations of our technological milieu; their voices reminding us that so much, but not everything, have been—and will be—lost.

Conclusion

In different ways, the works and practices discussed in this chapter attempt to preserve, evoke, and create connections with the extinct animal Other: to make them grievable, to refuse their disappearance, to bring them closer. In addition, many seek to remind audiences of humanity's role in recent extinctions, and prompt action lest more animals be lost. In the instance of "playback" efforts, the connection between sound and ecological impact is clear. In works where sonic connection is meant to provoke behavioral changes, their impact is trickier to measure. In this chapter, I argued that the collapsing of temporalities and invocation of presence afforded by sonic memorials to extinction create spaces of ongoing connection, refusal, and reflection. Here, deeper empathy—or, at least, uncomfortable intimacy—with lost beings, and nonhumans in general, may provoke audiences to action. Such connections and potential actions, however, do not occur in a vacuum. The political ecologies of field recording practices and the wider lives of audiences, for instance, are also entangled with how sonic memorials affect the mourning, resisting, and narrating of mass extinction. Even so, to give voice to extinct beings—either through recording, creating with, or speculating upon their sounds—is to recognize that they might have something to tell us. In the context of our current ecological horrors and their accompanying smorgasbord of denialism, backgrounding, and technological fixes, listening to lost species might matter more than ever.

This chapter did not offer an exhaustive overview of sonic memorials to extinction, even within the subcategory of those works and practices that "give voice"; for instance, Sally Ann McIntyre's *Huia Transcriptions* is a speculation upon extinct voices *in situ,* wherein a music box plays written notations of the species' song in their former habitats. In addition, Wolfgang Muller's *Séance Vocibus Avium,* an album containing speculative compositions of unrecorded extinct species, and other musical works that create with extinct creatures' voices, such as Lee Hyla's *Wilson Ivorybill,* were not covered. Indeed, this genre might also include sonic works that focus on endangered voices, like Maya Lin's *Sound Ring,* as well as historical animal sound speculation in the sciences (e.g. Senter). The fact that this chapter just scratched the surface of works and practices that reckon with extinction through sound only emphasizes the proliferation and significance of this genre. Undoubtedly, such sonic memorial efforts will only increase as more artists, museums, and scholars turn their attention to sound and, indeed, as more species fall silent.

Notes

1 A significant exception to this statement is "moments of silence", which feature prominently in public memorial culture, as do hymns and other musical compositions. In this chapter, however, I am more concerned with memorial objects (e.g. artworks, monuments, websites) rather than memorial events (e.g. funerals, wakes, commemorative days).

2 Though not the focus of this chapter, it is important to note that sound is not only a rhetorical resource for environmental activists, but an object of ecological concern. Sound pollution has been described as an invisible but catastrophic product of industrialization, drowning out birds' lines of communication, deafening whales, and negatively affecting human health. Moreover, some have considered sound *itself* as at risk: that "natural soundscapes" themselves are at risk of extinction (see, e.g., Krause; Hempton and Grossman)

3 These motivations sit within the context of various American conservation efforts in the twentieth century to resist the extinction of the country's native flora and fauna, including the passing of the Migratory Birds Convention Act in 1918 and the Endangered Species Act in 1973. For more detail on how the anxiety of species loss propelled the American environmental movement, see Barrow.

4 For instance, photographs of ivory-billed woodpeckers from the 1930s have also been seminal in debates about contemporary putative photographs of this species.

5 I use "humans" here to mirror the phrasing of the installation booklet. However, as many critical scholars have noted, it is not humans at-large, but rather it is *particular humans*, or *particular ways of being human,* that are to blame for our current ecological crisis (see, e.g., Wynter & McKittrick; Malm and Hornborg).

6 As with several other species represented in the project, Gould's Emerald sounds were never mechanically recorded, so the sculpture is based on like-species' songs. The exhibition catalog links to recordings of the Hispaniolan Emerald under this species (Turk).

7 For critical social science perspectives on de-extinction, see, e.g., Jørgensen; van Dooren and Rose.

Works Cited

Baiji, Yangtze River Dolphin. "Discovery of Sound in the Sea." Accessed 15 September 2022. https://dosits.org/galleries/audio-gallery/marine-mammals/toothed-whales/baiji-yangtze-river-dolphin/

Barclay, Leah. "Acoustic Ecology and Ecological Sound Art: Listening to Changing Ecosystems." *Sound, Media, Ecology*, edited by Milena Droumeva and Randolph Jordan, Palgrave Macmillan, Cham, 2019, pp. 153–177.

Barrow, Mark. *Nature's Ghosts.* University of Chicago Press, Chicago, IL and London, 2011.

Bastian, Michelle. "Whale Falls, Suspended Ground, and Extinctions Never Known." *Environmental Humanities*, vol. 12, no. 2, 2020, pp. 454–574.

Bezan, Sarah. "The Endling Taxidermy of Lonesome George: Iconographies of Extinction at the End of the Line." *Configurations*, vol. 27, no. 2, 2019, pp. 211–238.

Born, Georgina. "On Nonhuman Sound—Sound as Relation." *Sound Objects*, edited by James A. Steintrager and Rey Chow, Duke University Press, Durham, NC, 2019, pp. 185–207.

Boyer, Alison G. "Extinction Patterns in the Avifauna of the Hawaiian Islands." *Diversity and Distributions*, vol. 14, no. 3, 2008, pp. 509–517.

Brand, Albert R. *Letter to Peter Paul Kellogg,*Albert Rich Brand Papers, 21-18-899, 3 March 1936. Division of Rare and Manuscript Collections, Cornell University, Ithaca, NY.

Brand, Albert R., Albert A. Allen, Peter Paul Kellogg, James T. Tanner, and George M. Sutton. *"Hunting with a Mike" bound book of the Cornell University-American Museum of Natural History 1935 expedition*, n.d. Albert Rich Brand Papers, 21-18-899. Division of Rare and Manuscript Collections, Cornell University, Ithaca, NY.

Browning, Joseph. "Sound and More-than-Human Sociality in Catherine Clover's *Oh! Ah ah pree trra trra.*" *Organised Sound*, vol. 26, no. 2, 2021, pp. 179–189.

Bruyninckx, Joeri. "Sound Sterile: Making Scientific Field Recordings in Ornithology." *The Oxford Handbook of Sound Studies*, edited by Trevor Pinch and Karin Bijsterveld, Oxford University Press, New York, 2011, pp. 127–150.

Butler, Judith. *Precarious Life.* Verso, London and New York, 2004.

Carson, Rachel. *Silent Spring.* Houghton Mifflin, Boston, MA, 1964.

Comstock, Michelle and Mary E. Hocks. "The Sounds of Climate Change: Sonic Rhetoric in the Anthropocene, the Age of Human Impact." *Rhetoric Review*, vol. 35, no. 2, 2016, pp. 165–175.

Cortez, Alcina. "Museums as Sites for Displaying Sound Materials: A Five-Use Framework." *Sound Studies*, vol. 8, no. 1, 2021, pp. 43–72.

Craig, Wallace. "The Expressions of Emotion in the Pigeons. III. The Passenger Pigeon (Ectopistes migratorius Linn.)" *The Auk*, vol. 28, no. 4, 1911, pp. 408–427.

Debatty, Régine. "The Opera of Prehistoric Creatures." *We Make Money Not Art,* Accessed 22 April 2013. https://we-make-money-not-art.com/opera_for_prehistoric_creature/

DeLaure, Marilyn. "Performing Loss: Sonic Rhetoric in Maya Lin's *What Is Missing*?" *Liminalities*, vol. 15, no. 1, 2020, pp. 1–32.

Derrida, Jacques. "Above All, No Journalists!." *Religion and Media*, edited by Hent de Vries and Samuel Weber, Stanford University Press, Stanford, CA, 2001, pp. 56–93.

Dimensions in Testimony. "University of Southern California Shoah Foundation." Accessed 6 June 2022. https://sfi.usc.edu/dit

Eley, Craig. " 'Making them Talk': Animals, Sound, and Museums." *Antennae*, vol. 27, Winter, 2013, pp. 6–18.

Enright, Kelly. "Exhibiting Extinction: Martha and the Monument, Two Modes of Remembering Nature." *Cultural Studies Review*, vol. 29, no. 1, 2019, pp. 154–171.

Fuller, Errol. *The Passenger Pigeon.* Princeton University Press, Princeton, NJ, 2014.

Gallagher, Michael. "Field Recording and the Sounding of Spaces." *Environment and Planning D: Society and Space*, vol. 33, no. 3, 2015, pp. 560–576.

Gallagher, Michael. "Sound as Affect: Difference, Power and Spatiality." *Emotion, Space and Society*, vol. 20, 2016, pp. 42–48.

Gallagher, Tim. "The Birth of Natural Sound Recording." *Living Bird*, 15 Jan 2015. Accessed 6 June 2022. https://www.allaboutbirds.org/news/the-birth-of-natural-sound-recording/

Gallagher, Tim. *The Grail Bird: Hot on the Trail of the Ivory-billed Woodpecker.* 1st ed. Houghton Mifflin Company, Boston, MA, 2005.Gilmurray, Jonathan. *Ecology and Environmentalism in Contemporary Sound Art.* PhD thesis. University of the Arts London, 2018.

Gilmurray, Jonathan. "Sounding the Alarm: An Introduction to Ecological Sound Art." *Musicological Annual*, vol. 52, no. 2, 2016, pp. 71–84.

Greenberg, Cara. "Remarkably, Elizabeth Turk's Sculptures Highlight the Lost Voices of Extinct Birds." *Introspective Magazine*, 1 November 2020, Accessed 6 June 2022. https://www.1stdibs.com/introspective-magazine/elizabeth-turk/

Haraway, Donna J. "Teddy Bear Patriarchy: Taxidermy in the Garden of Eden, New York City, 1908–1936." *Social Text*, Winter, no. 11, 1984, pp. 20–64.

Hempton, Gordon and John Grossmann. *One Square Inch of Silence: One Man's Quest to Preserve Quiet.* Atria Books, New York, 2010.

Hunter, Hannah. "Digital Sound Archives can Bring Extinct Birds (Briefly) Back to Life." *The Conversation,* 25 February 2022, Accessed 6 June 2022. https://theconversation.com/digital-sound-archives-can-bring-extinct-birds-briefly-back-to-life-176115

Ibbotson, Rose. "De-extinction and Representation: Perspectives from Art History, Museology, and the Anthropocene." *International Review of Environmental History*, vol. 3, no. 1, 2017, pp. 21–42.

Jørgensen, Dolly. "Reintroduction and De-extinction." *BioScience*, vol. 63, no. 9, 2013, pp. 719–720.

Kanngieser, Anja. M. "Geopolitics and the Anthropocene: Five Propositions for Sound." *GeoHumanities*, vol. 1, no. 1, 2015, pp. 80–85.

"Kauai 'O'o". *YouTube,* uploaded by Robert Davis, Accessed 12 March 2009. https://www.youtube.com/watch?v=nDRY0CmcYNU

Kellogg, Peter Paul. *Letter to Clarence Cottam*, Peter Paul Kellogg Papers, 21-18-893, 23 February 1954. Division of Rare and Manuscript Collections, Cornell University, Ithaca, NY.

Krause, Bernie. *The Great Animal Orchestra.* Profile Books, London, 2012.

Malm, Andreas and Alf Hornborg. "The Geology of Mankind? A Critique of the Anthropocene Narrative." *The Anthropocene Review*, vol. 1, no. 1, 2014, pp. 62–69.

Marshall, Malaysia. "Designing RE: Climate Change. Episode 1: Steve Norton on Requiem." *Museum of Design Atlanta (MODA).* 2020. Accessed 6 June 2022. https://www.museumofdesign.org/steve-norton

Mosley, Imani Danielle. "Say Her Name: Invocation, Remembrance, and Gendered Trauma in Black Lives Matter." *Performing Commemoration,* edited by Annegret Fauser and Michael A. Figueroa, University of Michigan Press, Ann Arbor, MI, 2020, pp. 142–161.

Mundy, Rachel. *Animal Musicalities.* Wesleyan University Press, Middletown, CT, 2018.

Nixon, Lindsay. "Rebecca Belmore Wants Us to Listen to the Land." *Canadianart*, 7 June 2017, Accessed 6 June 2022. https://canadianart.ca/interviews/rebecca-belmore-landmarks-2017/

Norton, Steve. "R E Q U I E M, in memoriam twelve recently extinct species." *Requiem*, n.d. Accessed 6 June 2022. https://requiem.rednotebook.org/index.html

Oliver, Julian, and Crystelle Vu. "Extinction Gong." *Reckoning*, 31 January 2019, Accessed 6 June 2022. https://reckoning.press/extinction-gong/

Pinsky, Marin L., Anne Maria Eikeset, Douglas J. McCauley, Jonathan L. Payne and Jennifer M. Sunday. "Greater Vulnerability to Warming of Marine versus Terrestrial Ectotherms." *Nature*, vol. 569, 2018, pp. 108–111.

Reiley, Michael. "Echozoo: An Audio Portal to the Past and an Urgent Message to the Future." *Echozoo*, n.d. Accessed 6 June 2022. https://Echozoo.org/

Reynard, George. B. "The Rediscovery of the Puerto Rican Whip-poor-will." *The Living Bird*, vol. 1, 1962, pp. 51–60.
Ritts, Max. "Environmentalists Abide: Listening to Whale Music—1965–1985." *Environment and Planning D: Society and Space*, vol. 35, no. 6, 2017, pp. 1096–1114.
Robinson, Dylan. *Hungry Listening.* University of Minnesota Press, Minneapolis, 2020.
Rosenberg, Kenneth V., Adriaan M. Dokter, Peter J. Blancher, John R. Sauer, Adam C. Smith, Paul A. Smith, Jessica C. Stanton, Arvind Panjabi, Laura Helft, Michael Parr and Peter P. Marra. "Decline of the North American Avifauna." *Science*, vol. 266, no. 6461, 2019, pp. 120–124.
Ryan, James R. " 'Hunting with the Camera': Photography, Wildlife and Colonialism in Africa." *Animal Spaces, Beastly Places*, edited by Chris Philo and Chris Wilbert, Routledge, 2000, pp. 205–222.
Senter, Phil. "Voices of the Past: A Review of Paleozoic and Mesozoic Animal Sounds." *Historical Biology*, vol. 20, no. 1, 2008, pp. 255–287.
Stanescu, James. "Species Trouble: Judith Butler, Mourning, and the Precarious Lives of Animals." *Hypatia*, vol. 27, no. 3, 2012, pp. 567–582.
Steensen, Jakob Kudsk. "How a YouTube Video brought an Extinct Bird Back from the Dead." *Silica Magazine*, 25 May 2018. Accessed 6 June 2022. https://www.engadget.com/2018-05-25-kauai-oo-honeyeater-youtube-memorial-ghost-media.html
Stephens, Elizabeth. "Speculative Biology: Precarious Life in Art and Science Resurrection Projects." *Continuum*, vol. 34, no. 6, 2020, pp. 870–886.
Steutermann Rogers, Kim. "Wave of Hawaiian Bird Extinctions Stresses the Islands' Conservation Crisis." *Audubon,* 6 October 2021. Accessed 6 June 2022. https://www.audubon.org/news/wave-hawaiian-bird-extinctions-stresses-islands-conservation-crisis
The Sonic Memorial Project. "Picture Projects and Dotsperinch." Accessed 6 June 2022. https://sonicmemorial.org/
Tower of Voices. "National Park Service." Accessed 6 June 2022. https://www.nps.gov/flni/planyourvisit/tower-of-voices.htm
Turk, Elizabeth. *Tipping Point: Echoes of Extinction.* Hirschl & Adler Modern, 2020.
United States Fish and Wildlife Service (USFWS). "Proposed Rule: Endangered and Threatened Wildlife and Plants; Removal of 23 Extinct Species From the Lists of Endangered and Threatened Wildlife and Plants." *Federal Register*, vol. 88, no. 187, 30 September 2021, pp. 8–10.
van Dooren, Thom, and Deborah Bird Rose. "Keeping Faith with the Dead: Mourning and De-extinction." *Australian Zoologist*, vol. 38, no. 3, 2017, pp. 375–378.
Whitehouse, Andrew. "Listening to Birds in the Anthropocene: The Anxious Semiotics of Sound in a Human-Dominated World." *Environmental Humanities,* Vol. 6, 2015, pp. 53–71.
Wynter, Sylvia, and Katherine McKittrick. "Unparalleled Catastrophe for our Species? Or, to Give Humanness a Different Future: Conversations." *Sylvia Wynter: On Being Human as Praxis*, edited by Katherine McKittrick, Duke University Press, Durham, NC, 2015, pp. 9–89.
Young, Ashley, Kathayoon A. Khalil and Jim Wharton. "Empathy for Animals: A Review of the Existing Literature." *Curator*, vol. 61, no. 2, 2018, pp. 327–343.

4 Entangled Extinctions and Cultural Resurgence

Susanne Ferwerda

In the underwater kelp forests that surround the south-east coast of lutruwita/Tasmania, the tides are changing. Growing in depths up to forty metres under the sea surface, the semi-closed canopies of the giant kelp plants form a "key and iconic habitat that dominates many nearshore rocky coastlines in temperate and cold-water regions worldwide" (Butler et al. 2). Several kelp species grow in lutruwita/Tasmania, a region of islands south of the Australian continental mainland, including giant kelp (*Macrocystis pyrifera*) in the deepest coastal waters and bull kelp (*Durvillaea potatorum*) in the shallow edges at the shoreline, both providing crucial habitat for local marine life. But as these most southern Australian waters have started to increase in temperature with great acceleration, the kelp is quickly losing its footing.

Bleached corals and dying reefs have become well-known visible symbols to represent changing underwater Anthropocene environments. The surface water temperature off the coast of eastern lutruwita/Tasmania shows the "greatest rate of warming" (Shears and Bowen 1) of all ocean water in the Southwest Pacific, with an increase of +0.20°C decade^{-1} observed between 1946 and 2016 (Shears and Bowen 1), meaning a 1.4°C increase over the span of seventy years. It is in these warming waters that species such as the Tasmanian Spotted Handfish (*Brachionichthys hirsutus*) and Red Handfish (*Thymichthys politus*) fight against their looming extinction. Settler-Australian artist Lucienne Rickard currently works in nipaluna/Hobart, lutruwita/Tasmania. Over the course of her sixteen-month durational performance *Extinction Studies* on a podium in the Tasmanian Museum and Art Gallery (TMAG)—which Rickard has recommenced in 2022, now drawing solely Australian species—she drew a total of thirty-five recently declared extinct species from across the world with graphite on paper: reptiles, amphibians, plants, insects, birds, mammals, fish, worms and snails. The drawings are large—the dimensions of the sheet of paper were 2 × 1.5 m—and took days, weeks, sometimes months to draw, only to be erased as soon as the last line was drawn, yet leaving their traces superimposed. In late 2020, as the penultimate drawing of the first iteration of *Extinction Studies*, Rickard drew and erased the Smooth Handfish (*Sympterichthys unipennis*). Handfish live

DOI: 10.4324/9781003315957-5

in the rocky river bedding and kelp forests of the southern Australian coast, with several species endemic to south-east lutruwita/Tasmania only. Via the figure of the handfish, Rickard addresses the aftermath of the colonisation of Tasmania and its effects on current marine life.

The notion of extinction becomes more complex when analysed from the perspective of Tasmanian Aboriginal history and contemporary resurgence. Bull kelp has a long history of significance as the material used in lutruwita/ Tasmanian Aboriginal cultural and artistic objects such as baskets to carry water, and its potential loss has far-reaching cultural consequences. In the work of Palawa[1] Tasmanian Aboriginal artist Mandy Quadrio, kelp bridges ocean and shore, showing the entangled nature of the contemporary colonial sea frontier (Cameron ix). The history and proliferation of extinction myths in the Tasmanian context complicate the notion of mass or species extinction in this region. Decolonising these extinction narratives is paramount in undoing the harmful effects of extinction narratives in relation to Indigenous people.

Mandy Quadrio's kelp work critiques and unsettles the still prevalent narrative that Tasmanian Aboriginal people became "extinct" when Nuenonne woman Trukanini[2] died in 1876, when she was approximately seventy-three years old. Notions of race and genocide upset this harmful narrative that facilitated consistent attempts to erase Tasmanian Aboriginal cultures, despite their survival and resurgence. The bull kelp Quadrio uses in her work thus emphasises the self-determination of contemporary Tasmanian Aboriginal communities and reminds us that the survival of kelp is central to the revitalisation of material cultures, and vice versa.

Memorial culture and cultural memory practices are both harnessed and untangled in the work of Rickard and Quadrio. The mediation and remediation of extinction narratives (Bolter & Grusin) happens in the particular museum setting in which both these artists work, as well as in the reshaping of extinction *as* narrative. TMAG plays a crucial role in the afterlives of local colonial extinction narratives and myths, albeit with ethical consequences different from those produced by Rickard's and Quadrio's work. It is a significant extinction site because of its extensive Thylacine display and because of its extremely fraught history of displaying Tasmanian Aboriginal remains. The fallacies and enduring legacy of Tasmanian Aboriginal human extinction myths can be traced back to the display of Trukanini's remains between 1904 and 1947 at TMAG in an exhibition called "Truganini, The Last Tasmanian Aboriginal" and the racialised propaganda that followed her death in 1876. Drawing on TMAG as the site where Rickard and Quadrio's work is displayed, as well as teasing out their multispecies entanglements, I focus on concepts such as "erasure", "extinction" and "resurgence" to address extinction memorial culture and unfold the connection between contemporary representation of species extinction and the many lives that encounter kelp.

Lucienne Rickard: Multispecies Extinction and the Art of Erasing

For the duration of one month in late 2020, the image of a small extinct fish appeared and then vanished at TMAG in nipaluna/Hobart, lutruwita/Tasmania. The last and only Smooth Handfish (*Sympterichthys unipennis*) that was ever physically accounted for in Western records was the single fish scooped up with a dip net by French zoologist François Péron during a colonial expedition to south-eastern Tasmania in 1802 that was commissioned by Napoleon Bonaparte (Last and Gledhill 56). The handfish species are odd-looking but charismatic creatures. They prefer to use their "hands", their comparatively large pectoral fins, to crawl over the reef and river bedding. The single Smooth Handfish Péron captured, killed and preserved would become the holotype for the species, the single physical example to describe the species in its entirety. This little, 4.4 cm, single Smooth Handfish (Figure 4.1) was taken to the Museum National d'Histoire Naturelle in Paris, France, and described in the captain's journal as a little fish "which is unusual in that its foremost fins are exactly like hands" (Last and Gledhill 56). In the two centuries that followed, no other sighting of the Smooth Handfish was ever recorded even as, between February and June 2015, "nineteen experienced divers at twenty-two sites across southern Tasmania" went to "search for Ziebell's Handfish and the Red Handfish, which are presumably similar in ecology and distribution to the *S. unipennis*" (Last et al. "Sympterichthys unipennis"). Two hundred and twelve years after it showed itself to the colonial expeditioners just once, in March of 2020 the International Union for Conservation of Nature's (IUCN) Red List of Threatened Species declared the Smooth Handfish officially extinct.

Rickard sourced the species she featured in the first iteration of *Extinction Studies* from those declared extinct since the year 2000 on the IUCN Red List. What happens when extinct species reappear in Rickard's durational performance? What does it mean that all that remains of their appearance are faint traces on a sheet of paper, and how does this affect the memorial taking place on stage, as well as in their memory that lives on? The questions "what is lost?" and "what remains?" are present in Rickard's durational memorial performance, its precise museum location and the environments that these extinct species belonged to.

Erasure and reinscription are central to the symbolic and material aspects of Rickard's work. Rickard reuses the same piece of paper until its structural integrity can no longer facilitate her art. The particular species that Rickard draws encounter each other on paper as a palimpsest, creating new patterns and new relations. The graphite pencil and eraser marks become part of the paper, but the shavings pile up on the ground. They move when people walk past and shift the circulating air.[3] These erasures complicate the idea of extinction on both a local and global scale, end on an open note and call

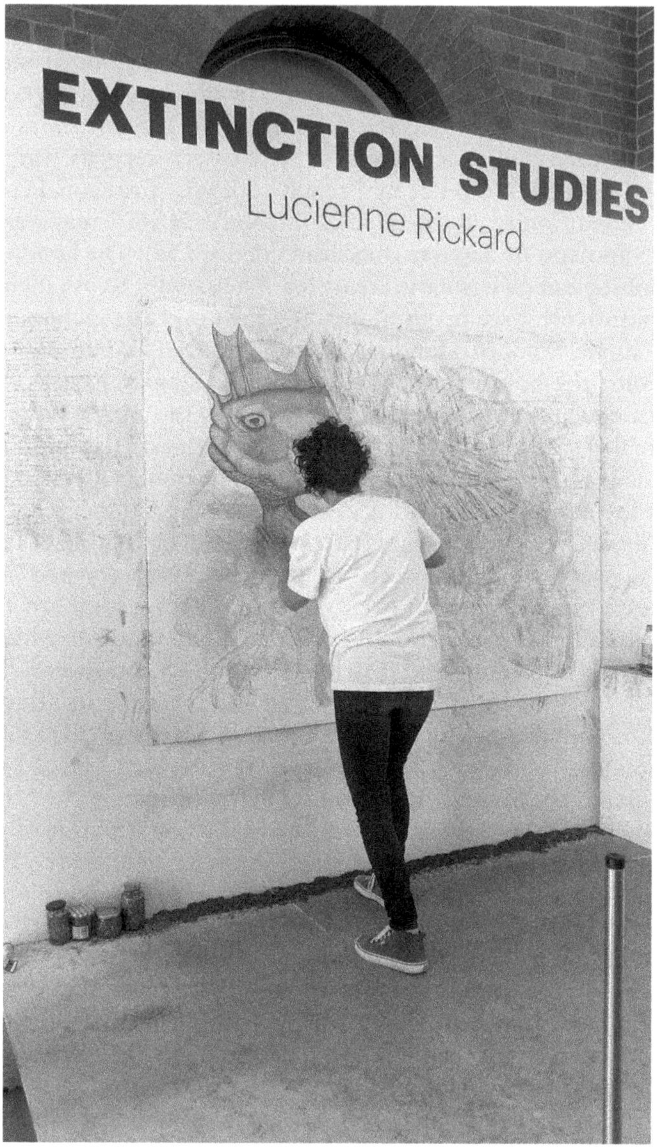

Figure 4.1 Rickard, Lucienne. *Extinction Studies.* Durational Performance: Graphite on Paper, 2 × 1.5 m, Exhibited at: Tasmanian Museum and Art Gallery (TMAG), Hobart, Tasmania, Australia, 9 December 2020. Author photo.

attention to the ongoing nature of extinctions. When Rickard erased the Smooth Handfish on 9 December 2020, this act not only emphasised their loss, but simultaneously amplified the plight of the three critically endangered Tasmanian handfish species that still crawl through the kelp forests and rocky reefs of south-eastern lutruwita/Tasmania. These endangered handfish, with their distinctive pectoral fins—the hands that they use to crawl through kelp forests and through river and reef bedding—are "charismatic minifauna" (Coates 274) and have become local environmental and conservation icons, alongside larger distinctive species such as the Tasmanian devil (*Sarcophilus harrisii*) and the Thylacine. Rickard's work performs and practises the remembrance of this fish that no longer swims lutruwita/Tasmanian waters but remains crucial to the survival of kindred species. By invoking species memories, her work asks us to imagine a future in which other species likewise could become only traces of themselves.

In *Extinction Studies*, Rickard draws the Smooth Handfish and, line by line, recreates the entanglement of touch and memory that precedes and follows her work. During the drawing stage of the performance, most time is spent between Rickard, her research on the species, their image, the pencil and the paper, with the occasional conversation with visitors of the museum. I attended several of the erasure events across the sixteen months of Rickard's performance. The affective nature of this project—performing, embodying and showing how extinction and erasure are connected—moved through the crowds that gathered to watch extinction happen in front of them. The entire crowd, including babies and toddlers, fell silent to watch Rickard's hands move over the page and superimpose erasure with the weight and responsibility we carry to think of extinction as more than any singular event. These moments where affect becomes collective are part of what Hannah Stark calls "extinction afterlives" (66). Diving deeper into the idea of extinction afterlives is helpful to further analyse how Rickard's work shifts the idea that extinction is a singular event, to the "longue durée" and "dull edge" (van Dooren 58) of colonial species extinction entanglement.

Extinction Afterlives and Their Underwater Colonial Entanglement

Two rooms away from Rickard's *Extinction Studies* performance at TMAG, the permanent exhibition "The Thylacine: Skinned, Stuffed, Pickled and Persecuted" shows the bones, skins and some of the only moving images of the Thylacine. The juxtaposition of these two exhibitions draws out the relationship between extinction, loss, mourning and the act of erasure (Ferwerda 2). The Thylacine is an extinct carnivorous marsupial, the last of which died in 1936 in the Beaumaris Zoological Gardens in nipaluna/Hobart, lutruwita/Tasmania after the species was hunted to death by graziers who believed they were livestock predators (Burbridge and Woinarski 2016).

The "extinction afterlives" and remnants of the Thylacine are displayed when "after death, they enter a museum collection" (Stark 66). These "uncanny" Thylacine remains, Stark notes, have

> something to teach us about how extinction happens and [...] reveal complex and interlocking stories about empire, the relationships between collectors, museums and zoos, the public desire to look at animals on display, and the individual lives, death and afterlives of particular animals.
>
> (77)

The permanent Thylacine exhibit performs what can be called the longue durée of their erasure. It shows how loss of a single species as the result of anthropogenic extinction events is part of the long and enduring history of settler colonialism, capitalism and Western notions of human exceptionalism. As Thom van Dooren argues in *Flight Ways*, humans are "implicated in the lives of disappearing others" (5) as individuals, as communities and as a species. This "dull edge" of extinction represents the "prolonged and ongoing *process* of change and loss that occurs across multiple registers and in multiple forms long before and well after [a] 'final' death" (58, italics original). Human-nonhuman extinction entanglements need to be taken seriously in order to account for the extrapolation of the notions of scale and time that are at stake at the edge of extinction. Extinction does not just happen when the last "specimen" of a species dies. As Stark and van Dooren show in their respective work on the afterlives and edges of extinction, extinction is a long event of unwinding patterns of multispecies relationships:

> To allow the term 'extinction' to stand for only the death of the last of a kind is to think with an impoverished notion of 'species', a notion that reduces species to specimens, reified representatives of a type in a museum of life, and in so doing ignores the entangled relations that *are* a particular form of life.
>
> (van Dooren 58)

Presented in juxtaposition with TMAG's permanent Thylacine exhibition, Rickard's *Extinction Studies* offers an alternative understanding of extinction events of the twentieth and twenty-first centuries. In Rickard's drawings, the act of erasing and the durational character of the project draw on important issues at stake in contemporary debates on the scale and time of what has become known as the "sixth" or the "Anthropocene mass extinction event" that has seen a large increase in extinction rates (Ripple et al. 2017). Growing awareness of the effects of anthropogenic planetary change means that hierarchical dualistic ideas such as human/nonhuman, nature/culture, so long at the core of Western thought, have slowly become unsettled. Positioning the human at the top of the species iceberg has obscured our ability to observe the devastating losses that have already occurred and are ongoing (van Dooren 18). As Timothy Clark notes in *Ecocriticism on the Edge* (2015): "the Anthropocene enacts the demand to think of human life at much broader scales of space and time, something which alters significantly the way that many once familiar issues appear" (13). The times and scales of extinction are broadened when species are no longer viewed in isolation but in touch with each other. This enables us to think of extinction as both "long" and "fast-paced": at once a "deep" and "shallow" process of layered multispecies change.

The TMAG Thylacine exhibit shows the same remains, day in, day out, in their own room separate from other long-term exhibitions. Down the hall, Rickard's drawings and erasures build up over time. The process of inscription and erasure on the same piece of paper emphasises both the fast-paced nature of contemporary mass extinction events, and the cumulative loss and building of grief that is invoked by gathering species that can no longer be "saved". *Extinction Studies'* cumulative nature moves away from the usual tracking of extinction via the loss and endangerment of single species, which Heather Swanson and her co-authors argue is not the best way to "see extinction":

> We often tally the plants and animals at risk of extinction one by one on lists of endangered species. But single species are not the best units through which to see extinction—because they are not the units of life.
> (Swanson et al. 141)

Life is made up of more complex patterns than these linear extinction lists suggest. Species loss does not just refer to the disappearance of a group of individual lives; it illustrates the loss of multispecies worlds. In *Extinction Studies,* Rickard rewrites the Red List and reframes the end of the world not as an apocalyptic event to come, but as a process set in motion long ago, from at least when Péron scooped the Smooth Handfish out of Tasmanian waters and took it to France. The slow unravelling of the world of the handfish is part of the long event of south-eastern lutruwita/Tasmanian coastal transformation, the disappearance of kelp environments and the changing nature of global underwater life, but it is also entangled with other extinctions in complex ways. Deborah Bird Rose notes that

> [r]ates of extinction are perhaps ten thousand times the background rate; as ecologist Steven Harding says, we are hemorrhaging species [...] numbers are a proxy to which it is worth paying attention. However, what is actually occurring is more dire than the numbers indicate.
> ("Shimmer" 52)

Reiterating Donna Haraway's words that "[i]t matters what stories make worlds, what worlds make stories" (*Staying* 12), when we look beyond the numbers, we can begin to see that "as things start to slip down that death road, other things start going too" (Rose, "Shimmer" 52).

The ocean has been a site of species loss for over two centuries. As a species that only visits underwater environments, it has been difficult for humans to assess the magnitude of marine bio- and ecodiversity loss. The grief and memorial culture, therefore, that follows the stories of terrestrial loss such as the last Dodo (*Raphus cucullatus*) or Passenger Pigeon (*Ectopistes migratorius*) has never, until recently, extended to the life that no longer lives underwater. When the Smooth Handfish disappeared in front of a human crowd at TMAG in December 2020, the audience was visibly and audibly affected. Witnessing the appearance and disappearance of species is confronting for a variety of reasons. From the perspective of the affect of grief and mourning, it offers a way into an alternative space and time of "acknowledgement and respect for the dead" (van Dooren 126) and a practice of memorial culture that emphasises both its collective and relational nature.

Museums are historically colonial institutions that "drove the growth of European empires" and "emerged as active tools of empire" throughout the eighteenth and nineteenth centuries (Giblin et al. 471). As museums have tended to produce only "narrow official histories" that included many instances of cultural erasure (5), a narrow idea of conservation and survival became enforced within Western museum discourse. Several Tasmanian handfish species have become part of this process of resurrection and regeneration. *The Handfish Conservation Project* includes a captive breeding programme for the Spotted Handfish in controlled tank environments. The handfish in this breeding programme form a control population in case something unexpected happens to wild handfish populations, so they can be bred to release in the wild or increase existing populations. There are, however, no guarantees when it comes to the laboratory or the museum. The human-centric history of the museum passes on an idea of care and involvement with extinct species that tends to halt at a static notion of remembrance and conservation or celebrates just singular species. Museums have only recently begun to acknowledge their contested position and their responsibility as contributors to species extinction (Cameron and Neilson). *Extinction Studies* plays with these conventions to unsettle the magnitude of extinction on the level of single species. While efforts to save the handfish that still crawl the bottom of the south-eastern coast of lutruwita/Tasmania are underway, the process of extinction is always already a multispecies, material and memorial affair. Ann Rigney notes:

> [i]t has become generally accepted that the arts are an important source of innovation in memory culture: not only because of their huge social reach and cultural longevity, but above all because of their ability to articulate stories that have not yet been told and bring them into circulation.
>
> (73)

Practising memorial culture precisely via the arts, the story of the handfish generates new memories for public environmental discourse.

The survival of the handfish does not depend on its conservation alone. If more kelp disappears, the water temperature keeps increasing, and the river bedding continues to be used for invasive aquaculture or boat moorings, handfish will inevitably continue to be affected. Taking the handfish out of the water into the laboratory is ineffective in isolation. Handfish-kelp-water-salt-ocean are part of the same material oceanic memory in south-eastern lutruwita/Tasmania. *Extinction Studies* goes beyond an immobile idea of either recovery or erasure and performs the multivocal nature of Anthropocene mass extinction events. The intricacy of Rickard's drawings demonstrates the dedication and care for the stories of the species she draws. As the eraser shavings pile up on the floor underneath Rickard's feet, the public are invited to grieve the loss of these species. This Smooth Handfish took more than a month to draw before it disappeared in a matter of minutes. The power of this performance, thus, lies in the way Rickard performs extinction *as* narrative, and makes the handfish memorable and more accessible to the imagination of her audience.

Mandy Quadrio: The Fiction of Extinction

The work of Palawa Tasmanian Aboriginal artist Mandy Quadrio rearticulates the problem of extinction in lutruwita/Tasmania and the legacy of misuse of the extinction myth to erase the survival of Tasmanian Aboriginal people. What happens to seeing extinction in lutruwita/Tasmania when these histories are foregrounded? Mandy Quadrio's Country—the term often used by Aboriginal peoples to describe their lands, waterways, seas, and the life connected with it, or, as Ngunnawal Elder Jude Barlow explains: "Country is everything. It's family, it's life, it's connection" (AIATSIS)—is the tebrakunna/Coastal Plains Nation in the north-east of lutruwita/Tasmania and the Oyster Bay Nation of eastern lutruwita/Tasmania. Quadrio uses a lot of materials in her work that are difficult to handle. The steel wool she uses is rough on the hands and rusts over time. The bull kelp is dense and leathery; it is tough but pliable when wet. Like the rusting of steel wool, kelp changes over time as it moves from one environment to the next. In her work, Quadrio connects materials from the lutruwita/Tasmanian environment, such as kelp and reeds, that transform by being reabsorbed into the ocean, to colonial materials like steel wool that in its encounter with the ocean becomes abrasive. The tensions between hard and soft, pliable and abrasive, of Country and imported with settler colonialism, that emerge in Quadrio's work, put the narratives of extinction that have been perpetuated in relation to Tasmanian Aboriginal people on edge. Quadrio's work is part of a wider cultural resurgence, evident in the work of other contemporary Tasmanian Aboriginal artists, such as Trawlwoolway artist Julie Gough and Trawlwoolway Elder and artist Lola Greeno. By untangling the complex

Figure 4.2 Quadrio, Mandy. *Here Lies Lies.* Installation. Resin. Exhibited at: Tasmanian Museum and Art Gallery (TMAG), Nipaluna/Hobart, Lutruwita/Tasmania, Australia, 15–23 November 2019. *Mandy Quadrio.* www.mandyquadrio.com.au/here-lies-lies/

materiality of past and present in settler colonial Tasmania, Quadrio's work upends the racial foundations of the Tasmanian Aboriginal extinction myth and strengthens the importance of Palawa Tasmanian Aboriginal sovereignty and self-determination.

Looking at extinction theory from the environmental humanities alongside historical research and the work of Tasmanian Aboriginal researchers who address the problem and fiction of extinction in relation to Tasmanian Aboriginal people, kelp connects land to sea, past to present and future, and renders connections between colonialism and extinction more complex than long considered. In Quadrio's bull kelp works from the *Here Lies Lies* (2019) installation in the Bond Store at TMAG, the fiction of extinction in relation to Tasmanian Aboriginal people is unsettled (Figure 4.2). How does the use of kelp disrupt the linear narrative of colonialism and extinction that was used against Tasmanian Aboriginal people? How can we reconsider the nature of erasure, so prominent in Lucienne Rickard's *Extinction Studies,* when foregrounding its use against the communities that have lived in lutruwita/Tasmania for tens of thousands of years before European invasion, up to the present?

Unsettling Tasmanian Extinction Myths

At the opening night of the 2019 "Hobiennale"—a portmanteau to describe the biennial art, music and performance festival in nipaluna/Hobart, lutruwita/Tasmania—there was no overlooking Mandy Quadrio's installation *Here Lies Lies* (2019). *Here Lies Lies* consists of five pieces, created by Quadrio specifically for the basement of the Bond Store Gallery at TMAG: a Tasmanian Aboriginal canoe overlaid with steel wool, a pillar of soft steel wool and three textual pieces that repeat the words "here lies lies" in bronze and resin in several places across the basement, at the entrance, before the sandstone wall in the back of the room, and jumbled in a pile of letters. The space was dark, with only the individual pieces illuminated. As Quadrio explains in conversation with Justine Youssef:

> [b]y illuminating only the individual works, I was able to use this colonial space to shine a bright light on the lies and fictions of Tasmanian history. It also has the suggestions of a crime scene. The historical and contemporary crime scene is consolidated by the colonialist moment work at the final moment of the installation, which appears as a large tombstone.
>
> (Youssef)

At the entrance to the basement, twelve bronze letters on the ground stop the viewer in their tracks. The shining words, "here lies lies", set the tone. The lies and crime scene Quadrio alludes to are the murder and forced removal of Aboriginal people in Tasmania since the early 1800s that led to the narrative of the so-called "extinction" of Tasmanian Aboriginal people, a harmful and persistent story that became engrained in Australian cultural memory and haunts Tasmanian Aboriginal communities to the present day.

As historian Rebe Taylor describes: "from the mid-nineteenth century, when other south-eastern Australian colonies marked the passing of the last 'full-blood' Aborigines of their 'tribes,' Truganini (*Trukanini/Trucanini*) was named the last of her 'race' " ("Genocide, Extinction" 407). Persistent ideas about "full-blood" Tasmanian Aboriginal people depended on the "now-obsolete ideas of race and blood: that humans could be classified into distinct racial types, and that the blood was the conveyor of those racial distinctions" (406). These convictions have resulted in both the erasure of contemporary Tasmanian Aboriginal communities and, on a national level in Australia, still inspire harmful political discussions about "blood quota" and contentious legal definitions that aim to define Aboriginality.

Arguing against the narrative of extinction of Tasmanian Aboriginal people is not to say that genocide did not take place. In the early decades of British colonisation, the so-called Black War and invocation of martial law between 1828 and 1830 allowed "soldiers to shoot, and settlers to capture, Aboriginal people" (Taylor 406). Thousands of Aboriginal people died.

Those who survived but failed to escape capture were forcibly moved and incarcerated in several settlements across the island region. About two hundred Aboriginal people were taken to Flinders Island, in the Bass Straight north of the lutruwita/Tasmanian main island. Only forty-seven people returned twelve years later in 1847 from the site named Wybalenna, when they were moved to the old convict station at putalina/Oyster Cove on the south-east coast of lutruwita/Tasmania, at that time called Van Diemen's Land (Taylor, *Unearthed* 140). The then and now famous Nuenonne woman Trukanini (*Truganini/Trucanini)* was one of the people held captive at the Oyster Cove station. Her life came to exemplify the horrendous treatment Aboriginal Tasmanians endured because of the expanding white settlement on the island. Trukanini (*Truganini/Trucanini*) accompanied Government Conciliator George August Robinson earlier in her life and had become increasingly known as the "last Tasmanian Aboriginal woman". Trawlwoolway artist Julie Gough emphasises that the Tasmanian Aboriginal extinction myth started well before her actual death:

> The public perception of and attention to Tasmanian Aboriginal people during the latter half of the 1800s was carefully directed at Trucanini, in Hobart. This woman was internationally seen as, and expected to act as, the last representative Tasmanian Aboriginal. Her life, well documented from 1830 until her death in 1876, brought her great attention and much isolation.
>
> ("Fibre Across Time" 18)

Through these persistent narratives, ideas concerning extinction and the murder and deaths of thousands of Tasmanian Aboriginal people became conflated. These stories silenced and erased "the parallel and continuing story of the women living on the islands of Bass Strait and on Kangaroo Island" (Taylor, *Unearthed* 140), as well as the many contemporary descendants of Fanny Cochrane Smith, who was born on Flinders Island in 1831 or 1832 and was denied "full-blood" status throughout her life.

After Trukanini (*Truganini/Trucanini)* died in 1876, she was initially buried at the Cascade Female Factory, but later exhumed. Her skeleton was put on display at the Royal Society of Tasmania, in the building that currently houses the Tasmanian Museum and Art Gallery. Trukanini (*Truganini/Trucanini*) had been acutely aware of what could happen to her body after she died. Having lived through the aftermath of William Lanne's death,[4] Trukanini (*Truganini/Trucanini*) is said to have begged to "be buried in the 'deepest part' of D'Entrecasteaux Channel, south of Hobart" (Taylor, "Genocide, Extinction" 409). Trukanini's (*Truganini/Trucanini*) remains were not returned to the Tasmanian Aboriginal community until 1976, and her ashes were then scattered in the D'Entrecasteaux Channel. It was not until 15 February 2021 that TMAG and the Royal Society of Tasmania formally apologised and took responsibility for their actions and role in exhuming, collecting and exploiting Tasmanian Aboriginal remains and cultural artefacts (Figure 4.3).[5]

Figure 4.3 Quadrio, Mandy. *Here Lies Lies.* Installation. Reeds and Steel Wool. Exhibited at: Tasmanian Museum and Art Gallery (TMAG), Nipaluna/Hobart, Lutruwita/Tasmania, Australia, 15–23 November 2019. *Mandy Quadrio.* www.mandyquadrio.com.au/here-lies-lies/

The influence of the Tasmanian Aboriginal extinction myth reaches well beyond the Australian continent. In the mid-1940s, it inspired the concept of "genocide", which in turn extended the myth and cultural memory of Tasmanian Aboriginal extinction. Taylor explains that the

> term genocide, meaning the killing of a people was coined in 1944 by Raphael Lemkin to define the crimes of the Holocaust. [...] Lemkin considered one of the clearest cases of historical genocide was the policies and actions in the 1820s and 1830s of the British in the colony of Tasmania.
>
> ("Genocide, Extinction" 405)

She notes that over time the myth of Tasmanian Aboriginal extinction and the classification of genocide had become blurred and become synonymous at a time when Australia, as a nation, was trying to come to terms with its past:

> Truganini and the history of Tasmania did indeed come to 'stand' [...] for all Australian colonial wrongs, an idea bolstered by a genuine belief that to mount the extinction of the Tasmanian Aborigines reflected a more evolved White Australia: a nation prepared to say 'sorry'.
>
> (410)

As *tebrakanna* country and Lee explain, these sentiments continued the erasure perpetrated by the colonial state: "I am a *trawlwulwuy* woman from *tebrakunna* country, north-east Tasmania, Australia, and my history as an Aboriginal Tasmanian is characterised by my non-existence. [...] We became the unspeaking non-beings against a backdrop of colonising excess and greed" (415).

It is in relation to this history that Mandy Quadrio critiques and reclaims TMAG and the Bond Store. The Bond Store is the oldest wing of TMAG and was built between 1824 and 1826 as a four-level brick building used to store dry goods such as grain, with the basement likely used to store tobacco and spirits. This makes the Bond Store one of the oldest remaining buildings built in the early years of the British colonisation of Tasmania, following the first colonial settlement in 1803, fifteen years after the "first fleet" colonised what is now called New South Wales (Taylor 406). By placing a Tasmanian Aboriginal canoe in this space, alongside the poignant statement that "here lies lies", Quadrio works to "expose the historic and continued denial of the existence of palawa people" (Youssef). The ongoing myth of Tasmanian Aboriginal extinction becomes upended in Quadrio's installation. She incorporates several Tasmanian watercrafts and emphasises the seafaring Tasmanian Aboriginal traditions through both the canoe in *Here Lies Lies* and her use of bull kelp in much of her other work. The sandstone memorial, accompanied by the words "here lies", emphasises the commemorative nature of the installation. The stones that make up the basement of the Bond Store were placed there at the same time as war was waged against Tasmanian Aboriginal Nations. "[W]hat bodies lie beneath the built environment" (Youssef), Quadrio asks. Her work not only reasserts the survival of Tasmanian Aboriginal communities, but also honours the ancestors lost in the past two hundred years, as well as the tens of thousands of years of history of what is considered one of the oldest living human cultures on Earth.

Carrying Water and Reimagining Extinction

In her work on death and extinction, Deborah Bird Rose discusses what she calls "double death": "the process that is driving the great unmaking of life in this era known as the Anthropocene" ("Multispecies Knots" 128). While death is integral to life on Earth, in the Anthropocene death has become amplified. Disjointed from the immanence of death that is necessary to life, Rose's notion of "double death" is related to the nonlinearity of death in the Anthropocene. Death and life are no longer necessarily entwined. Death follows death, unable to continue the process of renewal as ecosystems are "unable to recuperate their diversity. The death of resilience and renewal, at least for a while" (Rose, "Double Death"). The coloniality that underlies these double deaths is particularly visible on the Australian continent.

The arrival of European colonisers was a direct cause of Aboriginal deaths, and a death drive that extended to nonhuman animal and plant life on the continent and the seas that surround it. The devastation caused by colonisation is visible in the increasingly unstable bushfire seasons and loss of biodiversity. The deaths that are happening on a large-scale underwater in lutruwita/Tasmania—the kelp that is eaten by invasive sea urchins and the handfish that are unable to thrive in the increasingly barren coastal water—are entangled with this longer durée of double deaths, bearing the scars of the British colonisation of the region. The impacts of these deaths not only matter for the life that is to come, but they also cause the past to become undone. Time and scale converge in these Anthropocene death patterns. The amplification of death in the present is a problem of the future, the present and the past.

The bull kelp that Mandy Quadrio works with reinscribes the importance of following the ancestors in order to break away from the continuing process of colonialism-induced double death. Until recently, the double death of the ocean and marine life had gone largely undetected. In part spurred by the increased attention to the disappearance of Tasmanian handfish species, the problem of kelp forest loss has become visible on shore. In conversation with Justine Youssef, Quadrio says: "when I have been hunting and gathering with the aunties, I've noticed a decline in the bull kelp that inhabits the coastal shorelines" (Youssef). Tasmanian Aboriginal communities have only ever used small portions of kelp. The massive contemporary changes to kelp forest stability are not the result of "human interference" as such but are caused by particular human actions: by the colonisation and Westernisation of the region and the pollution, urbanisation and environmental change this has incited.

Seen from the perspective of Tasmanian Aboriginal extinction myths and the continued fight of Tasmanian Aboriginal communities to sustain agency over the narratives of their survival, the disappearance and potential extinction of local kelp become increasingly complex. Kelp loss is important on the wider scale of global climate change and environmental change set in motion by colonialism. In southern California, kelp forest regeneration is researched in relation to its potential to store CO_2 and possibly decrease ocean acidification and slow erosion (Rogers-Bennett and Catton 1–9). Moreover, in lutruwita/Tasmania, preventing further kelp forest loss has particular local urgency. In Aboriginal thought and definition of Country, kelp forests are connected, or rather, an intricate part of life in lutruwita/Tasmania that has been affected by the death patterns caused by European colonisation and settlement. The survival of bull kelp matters to the preservation, revival, and continuation of Tasmanian Aboriginal cultural practices and the reinscription of cultural memorial practices that no longer erase but celebrate and sustain Tasmanian Aboriginal cultures.

Quadrio's contemporary kelp works reflect the long history of bull kelp water carriers in Tasmanian Aboriginal culture as vessels to hold water and drink from. The bull kelp vessel she made for *The Climate Foundation* (Figure 4.4)—project that promotes "the rejuvenation of a thriving marine

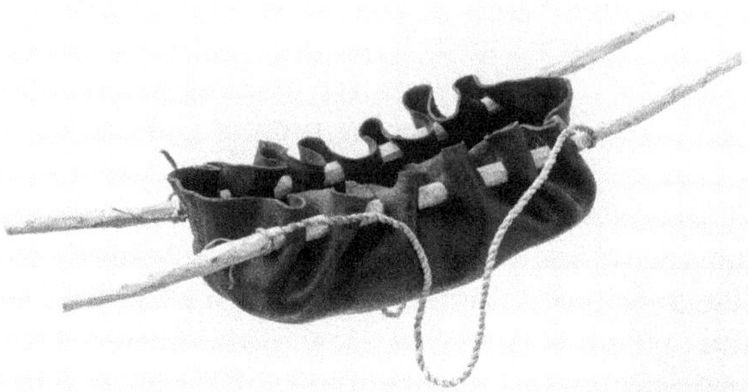

Figure 4.4 Mandy Quadrio. *Hold me while you can—Endangered species.* Bull Kelp, River Reed, Ti Tree, 2020. *Mandy Quadrio.* www.mandyquadrio.com.au/here-lies-lies/

environment, including bull kelp ecosystems and permaculture projects to regenerate life in the oceans" (Quadrio, *Hold Me While You Can*)—carries the name *Hold Me While You Can—Endangered Species* (2020). It is both a direct reference to the traditional water carriers and a continuation of this tradition, but with attention to climate change and the potential impact of kelp forest loss. Quadrio's kelp vessel comes from the same tradition as the oldest known surviving kelp water carrier, whose maker is unknown but likely came from the putalina/Oyster Cove settlement, south of nipaluna/Hobart. Commandant Joseph Milligan commissioned the water carrier for the 1851 Great Exhibition in the Crystal Palace in London, England, after which it was donated to the British Museum, where it still held today.

The short film *Tomalah* (2012) shows Julie Gough visiting the Tasmanian Aboriginal cultural objects held in the British Museum's storage facility. She writes:

> For one segment, I filmed myself unwrapping the oldest know Tasmanian Aboriginal kelp water carrier. [...] Recently in Tasmania I filmed and sound-recorded the kelp beds at low tide moving rhythmically at the southern end of Adventure Bay, Bruny Island. These kelp beds could well be the exact location where the kelp was collected to make

> the 1851 kelp carrier. […] I am linking the historic, lost overseas cultural object with the place it likely originated from, and the sounds of its long-lost homeland.
>
> (Gough, "Tense Past" 63)

The recording of Gough handling the more than one hundred seventy-year-old kelp carrier, interspersed with shots of contemporary lutruwita/Tasmanian bull kelp environments, demonstrates how important the continued survival of kelp environments is for the revival and continuation of Tasmanian Aboriginal traditions. In *Hold Me While You Can—Endangered Species*, Quadrio extends her earlier work to ask who will speak up for the kelp: "are we taking sufficient and appropriate actions to protect species that cannot speak for themselves?" (*Hold Me While You Can*).

Quadrio's kelp work emphasises the carrying and gathering materiality of these objects. Kelp is a gathering material, foundational to the structure of kelp forests. Cayne Layton and his co-authors call kelp a "habitat-forming species", or, in the industrialist language of the natural sciences, "ecosystem engineers" (2). Emphasising its gathering nature instead of the machinic, Quadrio and other Tasmanian Aboriginal artists, such as Vera Nichols, who work with traditional fibres, emphasise kelp baskets' gathering and storytelling nature. The baskets *are* material memorial culture; they do not just represent but also embody the thoughts and memories of Tasmanian Aboriginal women:

> The baskets are not empty. They are full of makers, their stories, their thoughts while making. The baskets are never empty. All of the thoughts jump out of the baskets onto all of us.
>
> (Nichols in Gorringe et al. iii)

Tasmanian Aboriginal women made fibre baskets out of kelp and several other plant fibres for a variety of reasons:

> [t]hey were strung around the neck and taken underwater to gather sea resources or carried on land while gathering or transporting goods. A basket's tenacity mirrored that of its makers. Diving underwater, climbing trees for possum or hunting seal, Aboriginal women made baskets for many purposes.
>
> (Gough, "Fibre Across Time" 3)

The resilience and tenacity of the fibres they used are echoed in these baskets and the memories they bring to life. Representing her culture on her own terms, Quadrio's kelp work redefines the discourse on extinction in lutruwita/Tasmania, not as a narrative of decline, but as one of resurgence and continuation: reinscribing memorial culture through artistic practice.

Conclusion

"Extinction is not just in the past", Dolly Jørgensen writes,

> it is very much in the present. That present then affects visions of what the future holds. A world of extinction is not a world without the species that have died; rather, it is a world in which those species change form. (213)

The extinction narratives that inform this chapter come from the same region in lutruwita/Tasmania. They have similar origins in the history of colonialism and invasive settlement by the British, and both include distinct elements of (attempted) erasure and a continued presence through cultural resurgence and embodied cultural memories. In Rickard's *Extinction Studies*, the colonial afterlives of species lost in recent times become present in the layering on the page. The disappearance of the Smooth Handfish from the environments it belonged to continues in the struggle for survival of other handfish species and the kelp forests they inhabit. The kelp encounters that happen on the lutruwita/Tasmanian coasts represent a local threshold of extinction. Kelp is under threat from global climate pressures, induced by colonial activities over the last several centuries, and the many lives that rely on these underwater forest habitats are feeling the consequences.

Practising memorial culture via the creative arts, kelp and handfish memories redefine the future of extinction *as* narrative. Kelp and handfish are entangled and inseparable from the extinction stories told about Tasmanian Aboriginal people, despite their continued survival and cultural resurgence. Kelp entangles past, present, future, sea and shore. The unique Palawa Tasmanian Aboriginal women's fibre art traditions continue and reiterate the importance of cultural memorial practices. As Quadrio's evocative kelp works show, recreating traditional baskets, kelp carriers and other contemporary versions of traditional crafts changes how we can see extinction. The importance of kelp survival is strengthened by the connections between land and sea, where that which touches kelp, between human and nonhuman, life above and under water, does not exist in the past, but always already takes place in the present.

Notes

1 *Palawa* is the *palawa kani* (the constructed Tasmanian Aboriginal language) word used to refer to today's Tasmanian Aboriginal community, "but *Pakana* the *palawa kani* word for 'people,' is also commonly employed as a label for the community" (Berk 12).

2 Trukanini's (*Truganini/Trucanini*) name has been spelled in several different ways over the last two hundred years. Nowadays, the common English spelling is *Truganini*. In *palawa kani*, her name is spelled as *Trukanini* or *Trucanini*.

3 Beyond the *Extinction Studies* performance, some records of the drawings remain. Photographs were taken by the public and by Rickard for her personal Instagram account. Moving images and photographs were also commissioned

by Detached Cultural Organisation, who funded the project and purchased the remains of the performance after it ended in January 2021. See for instance this short film by RUMMIN Productions on Vimeo: www.vimeo.com/393838689

4 When Tasmanian Aboriginal man William Lanne died in 1869, "local surgeons removed his head, hands, and feet" (Taylor, "Genocide, Extinction" 409). Lanne was a member of "the last free people captured in 1842" (Baird 16). Erroneously considered the "last full-blooded Aboriginal Tasmanian man", Lanne's mutilated remains were sent to museums and private collections across the world. William Crowther, who would become Premier of Tasmania, stole Lanne's skull and replaced it with that of another man. Lanne's hands and feet were later "found in the Royal Society rooms in Argyle Street, now TMAG, and a skull identified as Lanne's was finally repatriated from Edinburgh in 1991 and buried by the Tasmanian Aboriginal community in the north west of the state" (Baird 16). In 2022, the Hobart City Council voted to remove the prominent statue of Crowther from a park in the centre of Hobart.

5 For the apologies made on 15 February 2021, on behalf of the board of TMAG and the Royal Society of Tasmania, see: www.tmag.tas.gov.au/about_us/apology_to_tasmanian_ aboriginal_people, and www.rst.org.au/wp-content/uploads/2021/02/RST-2021-Apology-to-Tasmanian-Aboriginal-People-for-the-web.pdf

Works Cited

Baird, Andy. "Voices of Aboriginal Tasmania. *ningenneh tunapry* Education Guide." *ningenneh tunapry* Exhibition. *Tasmanian Museum and Art Gallery*, 2008, pp. 1–30. http://www.stors.tas.gov.au/au-7-0095-00738

Berk, Christopher D. "Palawa Kani and the Value of Language in Aboriginal Tasmania." *Oceania*, vol. 87, no. 1, 2017, pp. 2–20.

Bolter, Jay David and Richard Grusin. *Remediation: Understanding New Media.* MIT Press, Cambridge, MA and London, 2000.

Burbridge, Andrew and John Woinarski. *Thylacinus Cynocephalus, The IUCN Red List of Threatened Species 2016: e.T21866A21949291.* The IUCN Red List of Threatened Species, 2016.

Butler, Claire, Vanessa Lucieer, Simon Wotherspoon, and Craig Johnson. "Multi-Decadal Decline in Cover of Giant-Kelp Macrocystis Pyrifera at the Southern Limit of its Australian Range." *Marine Ecology Progress Series*, vol. 653, 2020, pp. 1–18.

Cameron, Fiona and Brett Neilson. *Climate Change and Museum Futures.* Routledge, New York, 2015.

Cameron, Patsy. *Grease and Ochre. The Blending of Two Cultures at the Colonial Sea Frontier.* Fullers Bookshop, 2011.

Clark, Timothy. *Ecocriticism on the Edge: The Anthropocene as a Threshold Concept*, Bloomsbury, London, 2015.

Coates, Peter. "Creatures Enshrined: Wild Animals as Bearers of Heritage." *Past and Present*, vol. 10, 2015, pp. 272–298.

Cuvier, Georges. *Mémoires du Muséum d'Histoire Naturelle*, vol. 3. Paris, 1817.

Dooren, Thom van. *Flight Ways: Life and Loss at the End of Extinction.* Columbia Press, New York, 2014.

Edgar, Graham, Rick Stuart-Smith, and Peter Last. *Brachionichthys hirsutus. The IUCN Red List of Threatened Species 2020: e.T2958A121210485.* The IUCN Red List of Threatened Species, 2020.

Ferwerda, Susanne. "Extinction and the Art of Erasing: Lucienne Rickard's Extinction Studies." *Art and Extinction,* special issue of *International Journal of Practice Based Humanities,* vol. 3, 2019, pp. 1–8.

Giblin, John, Imma Ramos, and Nikki Grout. "Dismantling the Master's House." *Third Text*, vol. 33, no. 4–5, 2019, pp. 471–486.

Gorringe, Jennie, Julie Gough and the Tasmanian Museum and Art Gallery (eds). *Tayenebe: Tasmanian Aboriginal Women's Fibre Work*. Tasmanian Museum and Art Gallery, 2009.

Gough, Julie. "Fibre Across Time: A History of Tasmanian Aboriginal Women's Fibre Work". *Tayenebe: Tasmanian Aboriginal Women's Fibre Work*, edited by Jennie Gorringe, Julie Gough and the Tasmanian Museum and Art Gallery, Tasmanian Museum and Art Gallery, 2009.

Gough, Julie. "Tense Past—Making My Way Back Home." *Fugitive History: the Art of Julie Gough*, edited by Julie Gough, UWA Publishing, 2018, pp. 3–65.

Gough, Julie. *Tomalah*, edited by Mark Kuilenburg. HDMI Video, Collection of the National Museum of Australia, 2015.

Haraway, Donna. *Staying with the Trouble: Making Kin in the Chthulucene*. Duke University Press, Durham, NC, 2016.

Jørgensen, Dolly. "Extinction and the End of Futures." *History and Theory,* vol. 61, no. 2, 2022, pp. 219–218.

Last, Peter and Daniel C. Gledhill. "A Revision of the Australian Handfishes (Lophiiformes: Brachionichthydae), with Descriptions of Three New Genera and Nine New Species." *Zootaxa*, vol. 2252, 2009, pp. 1–77.

Last, Peter, Graham Edgar, and Rick Stuart-Smith. *Sympterichthys unipennis. The IUCN Red List of Threatened Species 2020: e.T123423283A123424374*. The IUCN Red List of Threatened Species, 2020.

Layton, Cayne, et al. "Resilience and Stability of Kelp Forests: The Importance of Patch Dynamics and Environment-Engineer Feedbacks." *PLoS ONE*, vol. 14, no. 1, Public Library of Science, 2019.

Quadrio, Mandy. *Here Lies Lies*. Installation. Bronze, Steel Wool, Resin. Tasmanian Museum and Art Gallery (TMAG), Nipaluna/Hobart, Lutruwita/Tasmania, Australia, 15–23 November 2019.

Quadrio, Mandy. *Hold Me While You Can—Endangered Species*. Bull Kelp, River Reed, Ti Tree. Created for The Climate Foundation Project, 2020. www.mandyquadrio.com.au/kelp-culture-project/

Rickard, Lucienne. *Extinction Studies*. Durational Performance: Graphite on Paper, 2 × 1.5 m, Tasmanian Museum and Art Gallery (TMAG), Hobart, Tasmania, Australia, 6 September 2019–24 January 2021.

Rigney, Ann. "Cultural Memory Studies: Mediation, Narrative, and the Aesthetic." *Routledge International Handbook of Memory Studies*. Routledge, New York, 2015, pp. 65–76.

Ripple, William J., et al. " 'World Scientists' Warning to Humanity: A Second Notice," *BioScience*, vol. 67, issue 12, 2017, pp. 1026–1028.

Rogers-Bennett, Laura and Cynthia A. Catton. "Marine Heat Wave and Multiple Stressors Tip Bull Kelp Forest to Sea Urchin Barrens." *Scientific Reports*, vol. 9, no. 1, 2019, pp. 1–9.

Rose, Deborah Bird. *Double Death*. The Multispecies Salon, www.multispecies-salon.org/double-death/. Accessed on February 21, 2023

Rose, Deborah Bird. "Multispecies Knots of Ethical Time." *Environmental Philosophy*, vol. 9, no. 1, 2012, pp. 127–140.

Rose, Deborah Bird. "Shimmer. When All You Love is Being Trashed". *Arts of Living on a Damaged Planet. Ghosts of the Anthropocene*, edited by Anna Tsing et al., University of Minnesota Press, Minneapolis, 2017, pp. 51–63.

Shears, Nick and Melissa Bowen. "Half a Century of Coastal Temperature Records Reveal Complex Warming Trends in Western Boundary Currents." *Scientific Reports*, vol. 7, no. 1, 2017, pp. 1–9.

Stark, Hannah. "The Cultural Politics of Mourning in the Era of Mass Extinction: Thylacine Specimen P762". *Australian Humanities Review*, vol. 63, 2018, pp. 65–79.

Stuart-Smith, Jemina, et al. "Conservation Challenges for the Most Threatened Family of Marine Bony Fishes (Handfishes: Brachionichthyidae)." *Biological Conservation*, vol. 252, 2020, pp. 1–8.

Swanson, Heather, Anna Tsing, Nils Bubandt, and Elaine Gan. "At the Edge of Extinction". *Arts of Living on a Damaged Planet. Ghosts of the Anthropocene*, edited by Anna Tsing et al., University of Minnesota Press, Minneapolis, 2017, p. 141.

Tasmanian Museum & Art Gallery. *Tasmania: Earth and Life*, 2018. www.tmag.tas.gov.au/whats_on/exhibitions/permanent/tasmania_earth_and_life

Taylor, Rebe. "Genocide, Extinction and Aboriginal Self-Determination in Tasmanian Historiography." *History Compass*, vol. 11, no. 6, 2013, pp. 405–418.

Taylor, Rebe. *Unearthed: The Aboriginal Tasmanians of Kangaroo Island*. Wakefield Press, Adelaide, 2002.

Tebrakunna Country and Emma Lee. " 'Reset the Relationship': Decolonising Government to Increase Indigenous Benefit." *Cultural Geographies*, vol. 26, no. 4, 2019, pp. 415–434.

Youssef, Justine. *NAVA: In Conversation with Mandy Quadrio*. NAVA: National Association for the Visual Arts, December 2019. www.visualarts.net.au/podcasts/episode-49-mandy-quadrio/

5 Edenic Extinction

Memorialising Lost Species across Timescales at the Eden Portland Project

Clara de Massol

Introduction

In 2006, Sebastian Brooke, a British stone sculptor, started to carve figures of extinct species in stone. This was the beginning of the Mass Extinction Monitoring Observatory project (MEMO), a stone building to be built on the limestone cliffs of the Isle of Portland in Dorset, off the southwestern coast of the UK. The built spiral-shaped memorial was supposed to house these stone sculptures of lost species, operating as a monumental observatory of biodiversity. As the project developed, it was picked up by the Eden Trust, a charity and company running eco-tourism attractions in the UK and around the world. MEMO was absorbed by the Eden Trust to become Eden Portland, a cultural and natural heritage project that still plans to exhibit extinction, but this time underground, in a former industrial extractive site. The project is yet to be built, but when it is (and if it is, costs are estimated at £30 million), fossils and stone sculptures of extinct species will line underground galleries "where the ancient art of stone carving and the frontier technologies will breathe life into stories of biodiversity, extinction and evolution" (Eden Portland).

Eden Portland, then, is to be located in Bowers Quarry and Jordans Mine, right on the centre of the Isle of Portland. The site itself is an industrial and geological palimpsest, crossing the distance between human history and deep time. Portland is a central part of the Jurassic coast, the only British natural world heritage site, documenting 185 million years of geological history. The quarry's limestone, in turn, has been used for centuries to build the likes of St Paul's Cathedral in London and the UN Headquarters in New York. Today, the island, although it still contains an active quarry site, is full of empty mines; Eden Portland is due to (re)fill its void with the not-so-silent voices of extinction.

In this chapter, I examine the future site's narrative strategy and ideological discourses and how together they formulate a commemoration of biodiversity loss. To conduct such inquiry, the chapter focusses on the narratives formulated by the Eden Portland project by way of an analysis of the official communication and plans for the forthcoming extinction eco-attraction.[1]

DOI: 10.4324/9781003315957-6

Inquiries into the meaning-*makingness* of extinction often culminate in reflections of psychological, emotional and imaginative nature. To complement this, I propose here an examination into the ways in which extinction is, or can be, collectively remembered, with the aim to reflect on the ideological formation of extinction commemoration narratives.

The chapter is first concerned with the monumentality deployed in commemorating extinction. I then present Eden Portland as a layered and complex future site that combines deep geological history with the violent heritage of extractivism. This palimpsestic composition culminates in a neoliberal extinction narrative of human exceptionalism where I present Eden Portland's neoliberal politics of eco-tourism. The chapter ends with a reflection on necropolitics in the context of Anthropocenic extinction.

Eden Portland and the Eden Project

As mentioned, Eden Portland is part of the Eden Trust, a British charity dedicated to "collaboration in science, arts, technology and commerce", which seeks to create "a constituency for change and then putting it into action" (Eden Project). But while Eden Portland is yet to be built, the trust's civically engaged and socially minded ethos is carried through in the charity's main output, the Eden Project, a botanical garden, the world's largest greenhouse, and an eco-attraction and pedagogic centre located in Cornwall, approximately one hundred sixty km west of the Isle of Portland.[2] Built on the Eden Project environmentalist model, Eden Portland in turn looks to become an international environmental, cultural and touristic site, as well as a memorial destination; its overall mission aligning itself with the Eden Project and the Eden Trust's ideology of social change and responsibility.[3]

But as the flip side of the Eden Project, which looks to celebrate biodiversity and its potential, Eden Portland mediates and commemorates biodiversity loss. Both projects implement a social and environmental praxis by performing an anthropocentric and mission-driven ecological awareness. Howard Jones, former Organisational Development Director for the Eden Project, presents the educational and outreach principles of the site: "In the pace of the challenges that the twenty-first Century will bring, such as climate change, we need to build an awareness of how to sustain that which sustains us" (Jones 134). Alessio di Capua, the Eden Project's first (and former) Director of Philanthropy, continues along similar lines by presenting, in a 2018 interview, the charity's ambition to identify and support "projects around themes which are in Eden's DNA" (Cosgrove 14). While Eden's mission and strategic framework would have shifted in the ten years between the two interviews, the intent is to support and design projects that retain the core outreach and pedagogical ambition of the Eden Project, to build "relationships between people and the natural world" (Eden Project).

Figure 5.1 Cliff view of the Isle of Portland. Author photo.

The original Eden Project itself is located in what used to be a Kaolin clay quarry in Cornwall, UK. Respecting the project's "DNA", Eden Portland will similarly be built in a former stone quarry, this time east of the Eden Project, on the Isle of Portland in Dorset. The lineage between the two sites then exceeds their socio-environmentalist ambition; in terms of geological and material history, the two projects are inscribed within a long history of human extractivism and geological planetary formation, with Portland stone, a limestone formation of the Jurassic period, dating back about 200 to 145 million years ago, and the rock of Cornwall clay mines dating back about 420 million and 200 million years ago. And while the Cornish clay deposits began being exploited and extracted in the eighteenth century (Rowse 483), the stone of Portland has been quarried as early as the Roman Empire (Butler-Warke and Warke 960). Both projects are thus geologically grounded in a long history of deep time and extractivism. This heritage of extractivism will be explored further along in this chapter. For now, I want to focus on the site's deep time quality, and its relation to what is commonly referred to as the Anthropocene (Figure 5.1).

An extinction Memorial In and of the Anthropocene

Between educational attraction and eco-tourism, Eden Portland is a commemorative endeavour that seeks to *engrave* anthropogenic extinction onto pre-historic stones. As a memorial project, it situates our extinction moment as part of planetary heritage, in continuation with geological history, and with the Jurassic period, which saw the formation of the site's limestone foundation. This relates to the ways in which the "sixth extinction" (Kolbert 3), an ongoing extinction event ensuing from anthropogenic activity, is identified as central to an understanding of the Anthropocene. In that sense, Eden Portland, located within a significant geological site, would commemorate a phenomenon which itself is marking this new geological epoch of the Anthropocene, in which human activity has become the main geological force. Eden Portland is an Anthropocene memorial site and speaks to the epoch's – and its disturbances – deep time quality.

Here, the Anthropocene is a (contested) universal concept: if *we* – humans and nonhumans – are all in this (the Anthropocene), *we* do not experience it in similar ways. Some of us will become extinct, some of us will survive, and some of us will be mourned. Within Anthropocene pasts, presents and futures, there are differences in *grievability* (Butler) and in *memorability* (Rigney). But however significant, this is not per se the subject of this enquiry. Instead, the analysis looks to the commemorative narrative articulated by Eden Portland, what they formulate of extinction in Anthropocenic memorial terms.

While Anthropocenic extinction is widely discussed in scientific circles (Otto 4–5), its omnipresence in the public sphere is more recent. This growing awareness enables ecological narratives formulated by cultural and memorial institutions to lead the way in discussing extinction in public spaces from an Anthropocene perspective, and by doing so disrupt conventional Western perspectives of time, and of life and death, as contained in human terms. Discussing extinction from a geological perspective – something which the Eden Portland project aims to do – would potentially re-align the time of human, nonhuman and the planet together.[4]

The Anthropocene sheds light on the fact that human short-term actions have deep time consequences. This is significant from a memorial perspective since actions in the present have long-lasting effects on a future which might not exist for many human and nonhuman ecosystems. Extinction and its commemoration need to be central to any consideration of the Anthropocene. In other words, we need to look at what is lost and will be lost to comprehend the "shock of the Anthropocene" (Bonneuil and Fressoz), its "derangements" (Clark) and "trouble" (Haraway). This entails taking on a mnemonic task and designing commemorative practices that are adapted to the paradigmatic changes of the Anthropocene.

The aim of this chapter is to look at one of these commemorative endeavours, Eden Portland, and its infrastructure, the Eden Trust, to consider its narrative and what it proposes in terms of extinction remembrance. I argue that the efforts of Eden Portland steer away from a politically positioned narrative to instead propose a form of apolitical *gentle* environmentalism which does not make mention of the Anthropocene, nor its economic and political ramification. As opposed to this position (or lack thereof), I suggest that Anthropocenic extinction is deeply political because, as Rosine Kelz and Henrike Knappe affirm, "Mourning lost future possibilities, past and present losses of life or anticipated catastrophes are political practices that resist the dominant linear and development narrative in that they acknowledge the ruptures in time" (12). The manner in which ecological memorial practices are framed dictates the political potential of mourning and remembering in the Anthropocene.

I argue that where there is narrative there is positionality. In that sense, Eden Portland depoliticises extinction, in the same manner as "mainstream Anthropocene discourse [is being] widely criticized for turning a blind eye on power asymmetries and societal structures" (ibid.). So, while Eden Portland is not explicitly politically positioned, its apparent *neutrality* cannot be sustained. An illustration for that apparent lack of political positionality is the way in which, while proposing to "explore the importance of biodiversity" and "the threat of extinction" (Eden Portland), Eden Portland retraces an apolitical story of extinction and connects it to its site. Introducing the relevance of extinction ideas to its site, the Eden Portland website presents:

> In 1668, at the dawn of the Enlightenment, Robert Hooke stood on the construction site of the Royal Exchange, inspecting blocks of Portland stone. Hooke – sometimes referred to as 'London's Leonardo' – observed that the blocks were pitted and veined with strange geometries, concluding that these shapes must be the traces of creatures that had once inhabited the Earth, yet had subsequently vanished. This was the first realisation of extinction, and it was born from Portland stone.
>
> (Eden Portland)

This storying of extinction serves as a historical, geographic and geological contextualisation to the project, and neatly connects the subject-matter (extinction) to its site. But this fortuitous narrative is very much selective because, while the seventeenth-century "Scientific Revolution" in Europe, to which Robert Hooke was a figurehead, might have been the starting point of modern Western science, it was also deeply intertwined with the history of European imperialism (Harrison 56), something that is not mentioned anywhere in Eden Portland's official communication (Figure 5.2).

Imperial powers and their scientific discoveries owed much to colonial mechanisms of resource exploitation. Eden Portland's website mentions Robert Hooke and the Enlightenment but does not mention how Western scientific Enlightenment *discoveries* were then used by imperial powers to justify their so-called civilising missions, to bring knowledge and science to

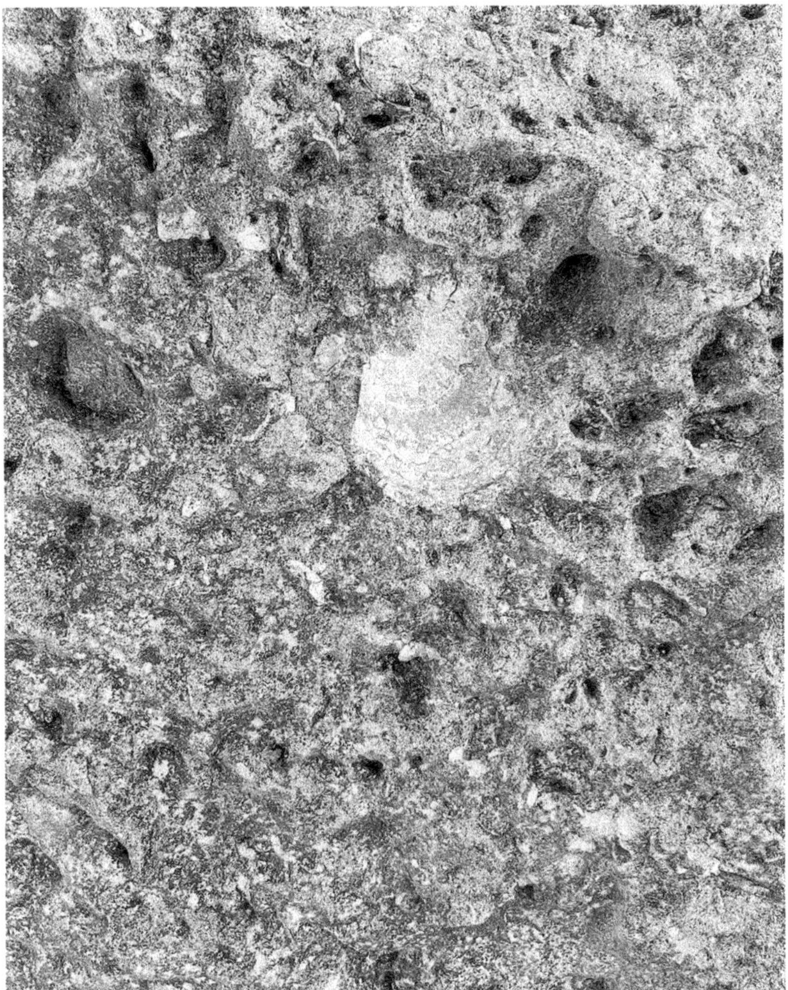

Figure 5.2 Portland stone. Author photo.

non-Western indigenous populations. This is also where the story of extinction comes into play. Biology in the seventeenth century was expanding in part because biological specimens would be brought back from colonies via, amongst other routes, the transatlantic slave trade (Drayton 17).

These alleged discoveries might never have seen the light of day without resources excavated from colonies, which then travelled across the globe and into Europe on slavery ships. In *Vanished: Hope and Histories of Extinction*, historian Sadiah Qureshi retraces the history of extinction and tells the story of French anatomist George Cuvier who, at the end of the nineteenth century, after having studied fossils of elephants that came to Paris from Asia and Africa, declared they were a different species. This discovery then supported a theory of extinction. Drawing from Qureshi's argument, it is thus argued here that a history of extinction, and consequently its memory and commemoration, is knotted with a history of European colonialism, imperialism and extractivism.

Extractive Monumentalism

From this perspective of exploitation and extraction and in memorial terms, the site of Eden Portland could be considered part of a *difficult heritage* (McDonald). McDonald's notion of difficult heritage looks to the legacy of painful histories, and how these, and their remnants, can – sometimes "difficultly" – become heritage, a part of collective history and identity. Beyond the fact that Anthropocenic extinction results from the brutal exploitation of the planet, its ecosystems and resources by petrocapitalist extractive industries, the very stones of Portland are testimonies to a violent and painful history.

Alice Butler-Warke and Matthew R. Warke, in their examination of the "heritage" significance of Portland stone, reveal that its "cultural value is based on characteristics that are largely unsavoury to a contemporary palate: convict labour, Crown seizure of land and resources, a use largely limited to 'elite' edifices, and a form of internal and international colonisation" (969).[5] While the celebration of Portland stone is centred around a skewed notion of British identity, it "is the same stone that has been the tangible manifestation of empire, of colonial domination, of an English-centric Union, and of an oppression of the working class" (ibid.). What this chapter adds to that argument is that the Eden Portland site would intensify the "difficult" aspect of the island's history in that in exhibiting extinction, it would unwittingly celebrate British extractive histories, themselves having contributed to Anthropocenic biodiversity loss.

If it were to become an intentionally difficult heritage memorial site, it would need to make visible the violence of extractive and exploitative industries, past, present and future, on human populations and cultures and on biodiversity.[6] By making present the absence of extinction in a public space through an attentive commemorative process, it could counter apolitical and ahistorical environmentalist narratives to devise a critical memorial politic of biodiversity loss.

As of yet, the project's website does not address any plans to discuss the site's violent extractive heritage and its relation to extractive economies. Still, as Butler-Warke and Warke argue, Portland stone belongs to a planetary heritage of exploitative brutality. Its presentation as British heritage is selective and is "being used to futureproof the capitalist transfer of the natural world into an elite commodity made possible through the labour of those who, as we have shown, are never supposed to enjoy the fruits of their labour" (969). Extractivism here refers to geological, political and economic processes of resource extraction and exploitation. More specifically, and as summarised by Thomas F. Purcell, "extractivism has emerged as a key concept to grapple with a whole host of politically urgent themes such as ecological destruction, global heating, accumulation by dispossession, depeasantization, land grabbing, and the financialization of nature" (1). Extractivism is thus directly linked to Anthropocenic extinction. There is potential for Eden Portland to explore and expose the deep heritage of the Anthropocenic extinction and its extractive lineage.

But as mentioned, from Eden Portland's initial plans, it does not seem like that is the direction the site and its director and trustees are taking. The Eden Portland project is not only designed as a memorial space, but also marketed as an "immersive" tourist attraction and *call to action*. Directly addressing its future visitor via an intimate "you", the project's website encourages prospective guests to visit the site: "Eden Portland invites you…" It continues further: "Using a provocative, playful and thoughtful fusion of science, art and storytelling the project will focus attention, engage emotions and inspire action". The site would make visible, albeit underground, biodiversity loss to make it appealing to the public. The idea of locating and containing an ecological memorial literally underground is significant. The importance of this particular subterrestrial location will be discussed further, but in symbolic terms, it is slightly ironic to claim to want to be a "catalyst" for change and "inspire action", all the while doing so buried under our feet. This is even more striking when faced with the challenges of representing the Anthropocene, with its scale, breadth and velocity exceeding human epistemological frameworks. It is emblematic of our myopic outlook to confine the significance of Anthropocenic extinction, its geological past and inhuman future, underground. The question is then: can visitors really be "inspired" and compelled to take "action" if this supposedly inspirational warning to our collective anticipatory extinction is contained in subterranean galleries?

Going back to our earlier argument, Eden Portland on its website does not make mention of extractivist and exploitative pasts, or of a desire to reckon with the violence of ecological loss and its connection to the British Empire. Instead, the project conveys a story of change, solutions and hope: it is about "focu[sing] attention, engag[ing] emotions and inspire[ing] action". It does claim to "explore the profound threats posed to modern biodiversity by the actions of humanity" but only to then "identify emerging solutions and innovations, and provide hope for a future that is safe, joyful and fulfilling". I argue then that this linear and monumental narrative of positivity and hopefulness emulates a modernist ideology of progress which conceals a reality and heritage of violent extractivism.

This narrative of hope and monumentality is exemplified in the plan for "the tree", a monumental stone carved Wood's Cycad tree that would sit inside Eden Portland's galleries. The Cycad tree is an emblematic species of extinction cultures: there is only one – noncloned – specimen left, and it can be found at the Durban Botanic Garden in South Africa. Eden Portland's sculpture is nine metres high and will sit in the mine. On its trunk, seven hundred "mason's mark" signatures from local schoolchildren will be carved. Beneath the mason's marks, visitors will be able to read Eden Portland's "call to action message", introduced by words from the project's (late) Royal Patron, Prince Philip: "This is our one speck of the universe". Of note here is the symbolism of carving the words of a monarchy (represented by the former consort of the British monarch) whose inherent purpose violently converges with past and present realities of exploitation and extraction on a memorial tree celebrating life. Besides, Prince Philip's words confirm the anthropocentrism at the core of the project: this "speck of the universe" is "ours". It is at the centre of the universe and carved in stone.

The figure of the tree is significant in memorial cultures. In their organic form, memorial trees are living, growing and standing witnesses to that which has been lost. Joy Sather-Wagstaff discusses the symbolism of trees in commemorative sites citing their resilience. The significance of trees in memorial cultures relates to their lifespan; they are placeholders for an absent past and for a slow and steady present, one of healing and reparation. They stand for the passing of time, the cyclicism of seasons and the perenniality of *nature* (Maddrell and Sidaway 131). They are active memory agents and carriers (Cloke and Pawson 107). Their usage in commemorative settings is about mimicking and "perform[ing] human time" (Heath-Kelly 71). So, in the manner of witness trees, like the forest of the martyr where 6 million trees were planted in West Jerusalem to commemorate Jewish Holocaust victims, or the memorial trees standing at the site of the World Trade Center attack in New York, the tree of Eden Portland stands witness to the absence of extinction. And while trees carry this notion of heritage and lineage, memorial trees in an extinction memorial would stand for the end of the line and answer for the end of what extinction represents: the finality of species lineage (Figure 5.3).

From an environmental memory perspective, trees have become, in the last decade, the figureheads of so-called sustainable memorial and funerary practices (Jeong). In a more optimistic light, trees are also the sign of a reckoning in memorial cultures that nonhumans, their form, ontology and temporality, need to be integrated into environmental memory discourses.[7] However, Eden Portland's tree – like the other future extinct species sculpture – is carved in stone, a traditional commemorative material, signalling ideas of monumentality, permanence and immutability. Eden Portland's tree is first and foremost a stone monument.

In order to consolidate the modernist ideology of progress mentioned previously, the site adopts the seemingly perennial form of stone monuments, exemplified by its monumental tree sculpture. In addition to the extractivism perspective outlined previously, the memorial politics of stone is significant here. The use of stone and concrete in monument-making can

Figure 5.3 Portland stone. Author photo.

seem old-fashioned to memory studies scholars and architects. From the 1960s onwards, monumental stone traditions have been rejected on the basis of their self-referentiality, with memorial politics steering away from the appearance of stability and foreverness and towards polysemy and mutability (Young). This is exemplified by conceptual artists Jochen Gerz and Esther Shalev Gerz's *Monument Against Fascism* which was built in Hamburg in 1986 as a twelve metres long steel-clad column which slowly sunk into the ground to disappear completely in 1993. Against the grain of what is becoming the standard in memorial discourse as a counter-monumental ambition, Eden Portland re-erects a stone monument, underground. It is thus proposed here that Eden Portland's usage of stones contributes to a refusal to reckon with Britain's brutal imperial pasts and presents. Instead, the site makes use of stones to tell grand historical narratives and thus instrumentalises the *longue durée* of Portland's Jurassic Stone Formation to create an exalting and monumental ecological experience. This results in a comfortable celebration of biodiversity, whereas extinction is at the service of an affirmative – and neoliberal – call to action.

However, stones also carry the potential for compelling Anthropocenic memorial narratives. From a geological perspective, stones are deep time witnesses of telluric changes. As Jeffrey Jerome Cohen exposes, they are "ancient allies in knowledge making" (3). Stones are, according to Cohen, agentive inhuman matter that transcends human time while accompanying our storying of the planet: they rend legible that which is beyond us. From this inhuman perspective, stones are an affective Anthropocenic memorial matter; they are both malleable (extinct species at Eden Portland are to be carved, or *set*, in stone) and impenetrable: "stone's time is not ours" as Cohen reminds us (11). If the (deep) historical complexity of Portland's stone were to be acknowledge and interpreted, it could potentially make for a *vibrant* carrier of Anthropocene memory.

Neoliberal Extinction

The commemorative strategy the site deploys is affirmative and individualistic: the website announces, "Eden Portland invites you to explore a subterranean labyrinth – a complex of mine tunnels that will be transformed into an unparalleled immersive space". The experience of the site is choreographed via the former mine's gallery, "a subterranean labyrinth" and "theatre of science". Visitors will penetrate the entrails of the earth, their journey paved by extractive industries: the site of both an explicit memorial to extinct species and an implicit monument to extractivism and to telluric excavation. I argue that through its commemoration of extinction, the site celebrates a form of neoliberal "environmentality".

Here, neoliberalism is understood beyond its market referent as having infiltrated and shaped individual subjectivities. Neoliberalism is a "political philosophy and ideology that affects every dimension of social life" (Giroux 52) in which subjects are "entrepreneurs of [them]selves" (Foucault 226). Neoliberal environmentalism relies on subjects to enact change through a form of internal governmentality. T.J Demos formulates this as the "neoliberalization of the Anthropocene" (To Save a World), a neo-humanist practice that licenses governments and corporations to deny realities of structural social and racial injustices, while asking of individual subjects to "save a world". This creates a paradoxical state in which subjects are asked to change their mode of consumption, but within already existing economic models, and by doing so to consume more, only slightly different products and services. This neoliberal environmentalism goes beyond injunctions to buy recycled products or switch to a fake meat diet, and it involves environmental cultural experiences or eco-tourism (Duffy 533), the like of Eden Portland. In that sense, Eden Portland is part of a wider phenomenon of "green economy" (Wanner 21), where the memoriality it formulates is stripped of its political potential and finds its apotheosis in the individual, not in the referential extinct species.

In his architectural study of the Eden Project, and relating it to neoliberalism, Ross Exo Adams discusses the site's presentation of time and space as in stasis. He proposes that the site formulates an "ecological time" as "a temporality of managed stasis" (177). This contributes to Eden Portland's depoliticised narrative of extinction. If *nature* is staged (at Eden Portland or at the Eden Project) as immutable, frozen in time and space, its presentation does not account for rapidly changing social conditions and its corresponding unstable environment. Related to this context, it is important to note the particular socio-economic position of the island of Portland. According to 2020 UK government data, Portland is among the 20% most socially and economically deprived areas in the country.[8] This situation is partly explained by the geographic isolation of Portland but mostly draws from the UK government's austerity measures and neoliberal-led welfare reforms.[9] The neoliberal depoliticisation of Eden Portland then does not exist in a vacuum but is inscribed within a socio-economic context: a neoliberal milieu that gives way to Eden Portland's individualist environmentalism.

As introduced previously, the curation of Eden Portland focusses on public advocacy and cultural change, the site being "a portal to an age of social, economic and ecological regeneration" (Eden Portland). The aim is thus to create the conditions for an active type of visit, as Tim Smit (co-creator of the Eden Project) promises: "This project will be the most innovative, exciting visitor experience in the world". The ambition is for visitors, young visitors especially, to witness the monumentality of extinction, and of Portland stone (and not the brutal monumentality of the British Empire) and get inspired to enact positive change. The emphasis is on possibilities – not loss – and on hope – not death. The overall mission is to empower visitors, showing them "light in the darkness" by "explor[ing] life and its interconnections". But beneath this hopeful narrative of "light in the darkness" is a more insidious neoliberal solutionism at play.

The site, in its inception, sustains a rhetorical framework of public advocacy, where collective effort and human ingenuity prevail: Eden Portland is presented as "a provocative, playful and thoughtful fusion of science, art and storytelling. [...] It will be a space for thinkers, scientists, educators and artists to interpret the story of life on Earth". But while doing so, it conceals the violent reality of extinction, only abstractly mentioning "the actions of humanity". This apolitical and ahistorical euphemism is symptomatic of a form of Foucauldian neoliberal "environmentality" (Agrawal 161; Fletcher 171), whereas the individual (read white and middle-class) can, or should, become the white saviour of the planet. Eden Portland's visitors, "inspired" to "act" (Eden Portland), are meant to become all-powerful entities, active force for good.

This form of environmentalism is not ahistorical: it is a product of solutionist missionary attitudes emerging from neoliberal individualist ideologies. The project itself has undertones of missionary work, with its pontifical, didactic and prophetic storytelling (unironically, it has named itself "Eden"); it calls on its prospective visitor to "[i]magine an underground cathedral [...] breath[ing] life into stories of biodiversity, extinction and evolution". The argument, drawing from Darwinian progress ideology, presents the (human) individual as an all-powerful godlike entity that can use its powers for good.

On the main page of Eden Portland's website, in big white letters, is announced: "We have read about biodiversity loss. Words have not been enough. This story must be touched, smelt, experienced and shared. It must catalyse celebration, reflection and action". The repetition of the imperative modal "must" underline the urgency and time-sensitive nature of the situation described, while the use of "we" at the start of the paragraph, the first word that is read, indicates a central anthropocentric subjectivity. Both these elements together produce a story of human exceptionalism, where "we" are the saviours and perpetrators of biodiversity stories. In the paragraphs below, it is explained that the site will allow visitors to understand "what it means to be human" and that "[t]ogether we will tell the biggest story of all" by getting "thinkers, scientists, educators and artists to interpret the story of life on Earth" to then "explore the profound threats posed to modern biodiversity by the actions of humanity". Humans are the sole "actors" and "interpreters" of the "story of life on Earth". The only mention of a nonhuman other is made through vague references to "biodiversity" and "extinction". It is subjected to the god-like "actions of humanity", which are also its only source of hope, a humanity that will "identify emerging solutions and innovations, and provide hope for a future that is safe, joyful and fulfilling". This universal humanity determines the biological makeup of the planet, good or bad. Extinction stories are at the service of humanity, used in the case of Eden Portland to help its visitor understand "what it means to be human". The narrative is one of essentialist humanism and is restricted by its anthropocentric terms.

As reported by John Blewitt in his research on the Eden Project, central to the site's mission is its visitors and their education. Blewitt interviews Jo Readman, Eden's Director of Education and Messaging, who, addressing past and future Eden visitors, declares: "you can make a difference. It is about celebrating and enjoying the environment you live in. Be positive about the future – you can do something" (177). This familiar "you", directly addressing the reader, is both generic – directed to *any* Eden visitors – and specific, focussed on this explicit individual that would visit the site and would have the illusionary power that this person can enact positive change.

And while Eden Portland will become a different site from the Eden Project, it carries out a similar mission to "catalyse celebration, reflection and action" (Eden Portland). The individual, this central "you" is the key to Eden Portland's extinction rhetoric. The future site is one of planetary heritage, but while it advocates for "ecological regeneration" (Eden Portland), it does so by iterating a linear narrative of progress, obscuring pasts and presents of injustice, extractivism and political inaction (Figure 5.4).

Figure 5.4 *Fallen Fossil* by Stephen Marsden, in Portland's Tout Quarry, a sculpture park and nature reserve located in one of the island's unused quarries. Author photo.

Conclusion: Unstable Necropolitics

The medium through which Eden Portland's narrative is mediated is a memorial site, a physical space where extinction will be "touched", "smelled", "experienced" and "shared". The site is dubbed a "theatre of science", "an underground cathedral", "a subterranean labyrinth" and "an unparalleled immersive space". What these descriptors share is their monumentality. The site will attempt to recreate the awe-inspiring experience of Portland's Jurassic coast to engage visitors with the monumentality of geological time and extinction. In the manner of the romantic sublime, Eden Portland is designed to convey strong physical emotions when faced with the monumentality of *nature*. But here the nonhuman is an instrument, the vector through which humans "catalyse celebration, reflection and action". This is symptomatic of an anthropocentric humanist memorial practice and an apolitical and ahistorical neoliberal ideology that depersonalises loss and decontextualises its source. To the visitor, extinction is not an environmental reality connected to anthropogenic disruptions like global warming and deforestation, but it is an experience and a missionary project. In this last section, I want to explore the inherent instability of the site: from the perspective of an *unstable* Anthropocene and because of its incompleteness (the site – at the time of the writing – has not being built yet).

From a public-facing standpoint, Eden Portland is designed to become a visitor-centred eco-attraction laying the (subterranean) foundation for a cultural commemoration of extinction. As expounded previously, it therefore could be examined as a difficult heritage site. Similarly, it could be studied from the perspective of thanatourism (death-related tourism): curating and exhibiting biodiversity loss and absences. In that sense, and without going into ethical considerations (by speculating on the ethical and moral limits of eco-thanatourism), Eden Portland is an example of environmental necropolitics. The analysis here deploys Nils Bubandt recontextualisation of necropolitics in, and of, the Anthropocene.[10] Anthropocenic necropolitics are conceptualised simultaneously in human and nonhuman terms. They describe the entangled interplay between death and ruination of human infrastructures and nonhuman entities, contaminated and contaminating that which is human and nonhuman. Bubandt explains: "In the Anthropocene, necropolitics operates under the sign of metaphysical indeterminacy rather than certainty, unintended consequences rather than control" (G125). Applying Anthropocenic necropolitics to Eden Portland would mean considering extinction within a "metaphysical [and temporal] indeterminacy"; the extinct species represented in the eco-heritage site operating across the human and the geological. Ultimately this indeterminacy, or instability, means that traces of absences are political. Bubandt clarifies: "In the necropolitics of the Anthropocene, geology is as entangled with politics as it is with ghosts. In the same movement that the Anthropocene is being established as a geological fact, geology itself is becoming political" (G136). I argue then that representing extinction (the traces – or ghosts – of what was) in an age of irresistible anthropogenic environmental disruptions (the Anthropocene) is a political matter.

This is even more relevant since, as of May 2021, Dorset Council – the local administrative agency governing the Isle of Portland – has approved an application for a million-tonne capacity landfill waste site to be located in Portland quarry. The temporal and "indeterminate" nature of waste (Hird 3) thus participates in the Anthropocenic necropolitics of Eden Portland and echoes the geological duration of its Jurassic stone structure. Anthropogenic waste, as epitomised by monumental landfills (it is estimated that it would take seventeen years for the Portland landfill to be full), here is considered from a posthuman perspective as exceeding human terms both in temporal and in ontological terms (Boetzkes; Zsuzsa and Lepawsky). Although

Eden Portland is not a landfill site, it is part of a landscape characterised by the monumental Anthropocenic processes of waste and extraction.

Anthropocene memory is neither comfortable nor linear. In mnemonic terms, the Anthropocene is unstable and cross-temporal. As Heather Davis and Etienne Turpin explain, being located within the temporal scale of the Anthropocene is "plac[ing] our industrialised present – a present that consumes time itself – within a temporal frame that is at once evolutionary and geologic" (6). The new epoch is a "threshold concept" (Clark *Ecocriticism*) in which the past, as the seeds of the future, conditions a present that is in process of becoming. Extinction is neither the end of the line, a prospective apocalypse, nor the sign of future redemptions, but the confirmation that collective memory needs to be conceptualised as co-constituted: extinction in the Anthropocene starts with the *Anthropos* but might not necessarily end with it. In place of this co-constituted extinction, Eden Portland formulates a tightly bowed narrative by turning biodiversity loss into what Jeffrey Olick terms a "usable past" that "serves the need of the present" (19). Extinction is formatted, mediated and exhibited to "serve" the pedagogic and inspirational function of the site, which is first and foremost a tourist attraction.[11]

However, extinction at Eden Portland is not yet set in stone, but it remains an idea. The site is there but for now is only inhabited by an imagined memorial to biodiversity loss. Like extinction in the Anthropocene, Eden Portland is unstable, yet to unfold, and still already conditioned by its context. While the memorial is not yet built, the quarry in which it will sit has been excavated for hundreds of years. The analysis – like its object – is bound to retain a speculative nature. The benefit of such prospective exploration is its openness: the approach is tentative so as to fit within an exploratory navigation into architectural speculation. Eden Portland might never be built after all, and even if it will, its narratives and curation are bound to change. But while this exploration is conjunctural (relying on unstable and changeable data), the argument remains systematic. It relies on a critical discourse analysis of the Project's official communication and takes on a constructionist position. This is inspired by Stuart Hall's socio-constructionist approach, whereas an object's discourse mediates, performs and constructs the knowledge it refers to. In other words, Eden Portland constructs knowledge on extinction and its commemoration. The analysis attends to that knowledge and *excavates* its meaning.

Anthropocenic extinction is of the past, of absences and losses, but it also of the future, of the species we are yet to lose. In terms of speculative biology and planetary heritage, the species that disappear today and that are yet to go extinct would have been the ancestors of tomorrow's biodiversity. But beyond a Darwinian logic, this rhetorical exercise can be interpreted as a call to decentre the human as the unique exceptional species. From a deep time perspective, human's cognitive abilities are only one amongst many other, past and future. Centring the human in extinction stories is doomed to fail since the myopic scale of modernist and Cartesian anthropocentrism does not leave space to account for the disappearance of nonhuman worlds, past, present and future.

Notes

1 Throughout the chapter, the analysis draws from the Eden Portland (referred to as "EP") project's official website.
2 The Eden Trust was initially created to support the Eden Project in 1996.
3 Although the trust supports Eden Portland, it is seeking additional funds from the government, private donors and its own investments to bring the project to life. And while there are no timelines available as to when the project will be realised, it is estimated that £30 million pounds are required for its realisation.
4 This is adopting a Western epistemological perspective. The Anthropocene is a Western concept, and non-Western cultures – aboriginal and indigenous notably – have conceptualised human time with the planet and vice versa. What used to be considered only indigenous cosmogonies is now being supposedly verified by Western science.
5 It is significant to note that Portland is home to HM Prison Portland, a male adult and young offender institution which opened as a prison in 1848. Butler-Warke and Warke (2021) refer to nineteenth-century prisoners who worked, and died, as labour convicts in Portland's quarries. This participates in the "difficulty" of Portland's history.
6 An example of a heritage site that intentionally engages with the difficulty of its past is the US National Memorial for Peace and Justice, informally known as the National Lynching Memorial, which is located in Montgomery, Alabama, near a former market site where enslaved African Americans were sold. The memorial consists of 805 hanging steel rectangles representing each of the US counties where a documented lynching took place. The memorial allows each county the possibility to take a second steel rectangle and install that marker in their own county. In that sense, it creates a visual marker where visitors can see which counties decided to reckon with their difficult past, and which did not. The site links a difficult history with the present and directly addresses racial violence in an effort to advocate for social justice.
7 Here, we can refer to Katie Paterson' *Future Library*, a forest planted north of Oslo, which will provide the paper for books to be published between 2014 and 2114. Another example is Rachel Sussman's *The Oldest Thing in the World* project, whereas the artist photographs the oldest living organism on the planet, amongst them a 9,550-year-old spruce tree in Sweden and an 80,000 -year-old colony of Aspen trees in Utah.
8 Here "deprivation" is characterised by a low life expectancy and social mobility, poor access to housing and essential services, low income and unemployment. See the Dorset Council 2020 *State of Dorset Deprivation* report.
9 See the 2019 UN report on "Extreme Poverty and Human Rights" in the UK documenting the damaging impact of welfare reforms and austerity policies (Alston, 2019).
10 In his research on the Indonesian mud volcano Lusi, which caused the displacement of some 39,700 people since 2006, Bubandt discusses the implication of necropolitics on the Anthropocene and vice versa. Lusi's 2006 eruption is said to have been caused by either a 6.3 magnitude earthquake or by nearby oil drilling: the life and death of the population around it subjected to both human and non-human irresistible forces.

11 This is assuming that Eden Portland will be modelled after the Eden Project, which brought over 1 million visitors in 2019.

Works Cited

Agrawal, Arun. "Environmentality: Community, Intimate Government, and the Making of Environmental Subjects in Kumaon, India." *Current Anthropology*, 46(2), 2005, pp. 161–190.

Alston, Philipp. "Report of the Special Rapporteur on extreme poverty and human rights to the Human Rights Council on his visit to the United Kingdom of Great Britain and Northern Ireland." *United Nations*, 2019. https://digitallibrary.un.org/record/3806308?ln=en.

Blewitt, John. "The Eden Project – Making a Connection." *Museum and Society*, 2004, 2(3), pp. 175–189.

Boetzkes, Amanda. *Plastic Capitalism: Contemporary Art and the Drive to Waste*. The MIT Press, Cambridge, MA, 2019.

Bonneuil, Christophe and Fressoz, Jean-Baptiste. *The Shock of the Anthropocene: The Earth, History, and Us*. Verso, London, 2016.

Bubandt, Nils. "Haunted Geologies: Spirits, Stones and the Necropolitics of the Anthropocene." *Arts of Living on a Damaged Planet*, edited by Anna Tsing et al. University of Minnesota Press, Minneapolis, 2017, pp. G121–G141.

Butler, Judith. *Precarious Life: The Powers of Mourning and Violence*. Verso, London and New York, 2004.

Butler-Warke, Alice and Warke, Matthew. "Foundation Stone of Empire: The Role of Portland Stone in 'heritage', Commemoration, and Identity." *Transactions – Institute of British Geographers*, 46(4), 2021, pp. 958–972.Clark, Timothy. "Derangements of Scale." *Telemorphosis: Theory in the Era of Climate Change*, edited by Tom Cohen. University of Michigan Press, Ann Arbor, MI, 2012, pp. 148–166.

Clark, Timothy. *Ecocriticism on the Edge: The Anthropocene as a Threshold Concept*. Bloomsbury, London, 2015.

Cloke, Paul and Pawson, Eric. "Memorial Trees and Treescape Memories." *Environment and Planning D: Society and Space*, 26(1), 2008, pp. 107–122.

Cohen, Jeffrey Jerome. *Stone: An Ecology of the Inhuman*. University of Minnesota Press, Minneapolis, 2015.

Cornwall Live. "Eden Project named top visitor attraction in Devon and Cornwall." *Cornwall Live*, 16 July 2021. https://www.cornwalllive.com/news/cornwall-news/tourist-attractions-devon-cornwall-among-4328391

Cosgrove, Sarah. "Interview: The Eden Project's New Director of Philanthropy: New Director of Philanthropy Points to 'Untapped Potential' to Raise Funds for Specific Areas of The Eden Project's Work." *Horticulture Week*, September 1, 2018, p14.

Davis, Heather and Turpin, Etienne. *Art in the Anthropocene: Encounters Among Aesthetics, Politics, Environments and Epistemologies*. Open Humanities Press, London, 2015.

Demos, T. J. "To Save a World: Geoengineering, Conflictual Futurisms, and the Unthinkable." *E-Flux Journal*, October 2018. https://www.e-flux.com/journal/94/221148/to-save-a-world-geoengineering-conflictual-futurisms-and-the-unthinkable/

Dorset Council. *State of Dorset Deprivation*, 2020 https://mapping.dorsetcouncil.gov.uk/statistics-and-insights/topics/Topic/Deprivation

Drayton, Richard. *Nature's Government Science, British Imperialism and the Improvement of the World.* Yale University Press, New Haven, CT, 2000.

Duffy, Rosaleen. "Nature-Based Tourism and Neoliberalism: Concealing Contradictions." *Tourism Geographies*, 17(4), 2015, pp. 529–543.

Eden Portland. *Eden Portland,* 2022. https://www.edenportland.org

Eden Project. *Eden's Mission*, 2022. https://www.edenproject.com/mission

Eden Project. *Eden Structure and Team*, 2022. https://www.edenproject.com/about-us/eden-structure-and-team

Fletcher, Robert. "Neoliberal Environmentality: Towards a Poststructuralist Political Ecology of the Conservation Debate." *Conservation and Society*, 8(3), 2010, pp. 171–181.

Foucault, Michel. *Naissance de la Biopolitique.* Gallimard/Editions du Seuil, 2004.

Giroux, Henry A. *Proto-fascism in America : Neoliberalism and the Demise of Democracy.* Phi Delta Kappa Educational Foundation, 2004.

Hall, Stuart. *Representation: Cultural Representation and Signifying Practices.* Sage Publication, Los Angeles, CA, 1997.

Haraway, Donna. *Staying with the Trouble: Making Kin in the Chthulucene.* Duke University Press, Durham, NC,2016.

Harrison, Mark. "Science and the British Empire." *Isis.* [Online] 96(1), 2005, pp. 56–63.

Heath-Kelly, Charlotte. "Survivor Trees and Memorial Groves: Vegetal Commemoration of Victims of Terrorism in Europe and the United States." *Political Geography* 64, 2018, pp. 63–72.

Hird, Myra J. "The Phenomenon of Waste-World-Making." *Hyperrhiz: New Media Cultures*, 30, 2016, p. 3.

Jeong, Saimi. "Pushing up trees: Is natural burial the answer to crowded cemeteries?" *The Guardian*, 10 November 2018. https://www.theguardian.com/lifeandstyle/2018/nov/11/pushing-up-trees-is-natural-burial-the-answer-to-crowded-cemeteries

Jones, Howard. "The Eden Project: A Synopsis around Sustainable Enterprise." *The Journal of Corporate Citizenship*, 30, 2008, pp. 133–137.

Kelz, Rosine and Knappe, Henrike. "Politics of Time and Mourning in the Anthropocene." *Social Sciences*, 10, 2021, pp. 12–15.

Kolbert, Elizabeth. *The Sixth Extinction: An Unnatural History.* Henry Holt and Company, New York, 2014.

"Landfill site agreed near Dorset World Heritage coast in Portland." *BBC*, 30 May 2021. https://www.bbc.co.uk/news/uk-england-berkshire-61632069Maddrell, Avril, and Sidaway, James, D. *Deathscapes: Spaces for Death, Dying, Mourning and Remembrance.* Ashgate, 2010.

McDonald, Susan. *Difficult Heritage. Negotiating the Nazi Past in Nuremberg and Beyond.* Routledge, London, 2018.

Olick, Jeffrey, K. "From Usable Pasts to the Return of the Repressed." *The Hedgehog Review*, 9(2), 2007, pp. 19–31.

Otto, Sarah P. "Adaptation, Speciation and Extinction in the Anthropocene." *Proceedings of the Royal Society B*, 285, 2018, pp. 4–5.

Purcell, Thomas, F. "Contesting Total Extractivism: Between Idealism and Materialism." *Journal of Agrarian Change*, 22(3), 2021, pp. 1–6.

Qureshi, Sadiah. "Vanished: Hope and Histories of Extinction." University of Reading, History Department Annual Stenton Lecture, 18 November 2021.

Rigney, Ann. "Differential Memorability and Transnational Activism: Bloody Sunday, 1887–2016." *Australian Humanities Review*, 59, 2016, pp.77–95.

Rowse, Alfred, Leslie. "The Cornish China-Clay Industry." *History Today*, 17(7), 1967, p. 483.

Sather-Wagstaff, Joy. "Trees as Reappropriated Heritage in Popular Cultures of Memorialization: The Rhetoric of Resilient (Human) Nature." *Encounters with Popular Pasts, Cultural Heritage and Popular Culture*, edited by Helaine Silverman and Mike Robinson Springer, 2015, pp. 235–250.

Wanner, Thomas. "The New 'Passive Revolution' of the Green Economy and Growth Discourse: Maintaining the 'Sustainable Development' of Neoliberal Capitalism." *New Political Economy*, 20(1), 2014, pp. 21–41.

Young, James E. "Memory and the Monument after 9/11." *The Future of Memory*, edited by Rick Crownshaw et al. Berghahn Books, New York and Oxford, 2010.

Yusoff, Kathryn. *A Billion Black Anthropocenes or None*. University of Minnesota Press, Minneapolis, 2018.

Zsuzsa, Gille, and Lepawsky, Josh. *The Routledge Handbook of Waste Studies*. Routledge, London, 2021.

6 Franklinia in the Garden

Memorializing Foliage, Preserving Heritage

Kelly Enright

Along a quiet rural road in southeastern Georgia lies an historic marker to the first species Anglo-Americans recorded disappearing from the landscape, a small tree with lavish white blooms named *Franklinia alatamaha*. Eighteenth-century naturalists and plant collectors observed just a few swampy acres of the species near the Altamaha River before its last sighting in 1803. Many looked for it in the century that followed, imagining rediscovery, but the species remained absent from its known habitat.

Yet, Franklinia itself was not entirely lost. Naturalists John and William Bartram collected and propagated the tree in their Pennsylvania garden, and shipped specimens abroad. Their efforts saved the species from complete extinction, earning it instead the IUCN designation "extinct in the wild." Franklinia grow today in botanical gardens and arboretum, backyards and public parks, and cemeteries and laboratories. Attempts to reintroduce it to its Altamaha habitat have failed. For more than two hundred years, *Franklinia alatamaha* has grown only in propagation.

Casting an historical lens on the history of this plant interrogates what it means when a species is extinct in the wild and the interplay between past and present inherent in that phrase. "The wild" is an imprecise location, and as environmental humanities scholars have argued, the concept of wilderness ignores historical change, whether that be geological, ecological, or cultural.[1] Extinction studies is increasingly contemplating the meaning of wildness as it relates to extinct species that appear to retain an imagined sense of wildness even after death.[2] While the term wild implies a lack of change, extinct implies a radical one. When combined, the two suggest an incomplete narrative. The phrase "extinct in the wild" becomes itself a kind of memorialization of a species' past life that ignores, or at least makes secondary, its current survival and the spaces where it blooms.

The history of Franklinia is a narrative of loss and gain. Franklinia may be extinct in the wild, but it is saved in the garden. This essay will examine the Georgia marker, a traditionally wrought site of remembrance associated with death and loss, then move to the gardens where Franklinia continue to live to argue that this too is a kind of memorialization. In her essay on seed banks, Anna-Katharina Laboissière writes that gardens "developed

DOI: 10.4324/9781003315957-7

as miniature representations of a harmonious political and cosmological order" (67). Gardeners are ever-imposing order upon the landscape, making and remaking the meanings of each plant within this new aesthetic, commercial, and domesticated context. From the time Bartram planted Franklinia to today, people have made Franklinia an artifact and souvenir of an imagined colonial past. While many species are saved because they are charismatically cute, Franklinia has continued to exist because of its associations with national and horticultural heritage.

In Situ Remembrance

The Georgia monument is a thick, white and gray rectangle of granite atop a short pedestal, with a beveled edge as its only embellishment, resembling nothing more than a tombstone (Figure 6.1).

Etched on the surface at top is FRANKLINIA ALATAMAHA in all caps. Inscribed below is a narrative of its history, bordered at bottom by the names of the organizations involved in placing the marker: the Long County Garden Club, Georgia's Division of State Parks, Historic Sites & Monuments, and the Department of Natural Resources. In the distance, the riverbank is lined with native pines, but the nearest tree is a crepe myrtle—a non-native species—seemingly planted to help complement the marker in its rather unceremonious placement along the road.

It was placed here in 1941 to celebrate the 176th anniversary of Bartram's description of Franklinia, the first known identification of the species to western science. Those who most wished to remember Franklinia in its native habitat were the members of the local garden club who seem to have crafted this monument to extinction to make connections between local and national heritages. The marker suggests the plant lives in memory because of its mystery, the text calling it "famous as The Lost Franklinia," a species no longer "growing wild." It points no finger and places no blame for the loss. It concludes referencing the Bartrams' saving of the species, stating that the tree that grew in their garden "became the ancestors of the present cultivated plants," invoking a sense of loss on the landscape but pointing to a living legacy.

The causes of Franklinia's in situ extinction are complex. After Bartram, it was collected as a garden curiosity, undoubtedly placing strain on the already small number of trees; nearby agriculture and imports of the Columbian Exchange likely also stressed the population. But it seems to have been a Pleistocene species already on its way out when Europeans encountered it. Charles Lyell spent time thinking about its extinction when he visited the region, contemplating why this species would suffer and not others, like the rather similar *Gordonia lasianthus* (Lyell 259–260). In much of the writing about Franklinia, even in the twenty-first century, uncertainty thrives, lending an aura of mystery to the acts of remembrance in which people engage.

Figure 6.1 The stone marker erected to honor Franklinia near the Altamaha River in Georgia. Photo credit: Michael Rivera.

The Georgia monument deploys the language of loss clearly in both text and form. It is not unlike a monument to an individual person, using the form of mourning most familiar to evoke an emotional response. Monuments are important cultural spaces of remembering that communicate shared values and identities.[3] It seems resigned to the garden existence of the species and proud of its connection to colonial and horticultural history. Invoking the names of well-known botanists and founding father Benjamin Franklin—the honoree of the genus name—creates a sense of shared heritage between rural Georgia and larger national history.[4] The species name, alatamaha (a Timucua word, variously spelled Alatamaha, Altamaha, Altama), refers to the river, vilified in a 1770 poem by Oliver Goldsmith as a place of extreme and dangerous wildness that was luring Brits away from their native home, connecting the species to the pre-contact landscape and a heritage of perceived wildness.[5]

The language of monuments often seems simple and straightforward, but they are powerful vehicles for communicating collective heritage that can be simultaneously inclusive and exclusive. The Georgia monument reveals that local people at the time cared about the lost plant, but not in a way that inspired conservation or changes to how they interacted with nature. The connection to domesticated Franklinia erased despair and the need for action, while connecting the landscape to a larger vision of American heritage that Franklinia represents. In the early twenty-first century, The Nature Conservancy attempted to reintroduce Franklinia to its Altamaha habitat, but both the soil and the plant have changed in the two hundred years since its extinction and all attempts have failed.

Ex Situ Artifact

In 1947, just a few years after the Georgia monument was placed, the city of Philadelphia planned to preserve Independence Hall, a colonial structure important as the site of the signing of both the Declaration of Independence and the U.S. Constitution. The plan included the grounds surrounding the building, in what became Independence National Historic Park, preserving several Revolutionary-era buildings, including Ben Franklin's home. The landscaping was meant to honor this same past, and botanist Charles F. Jenkins, then running the Hemlock Arboretum, wrote an enthusiastic and detailed letter advocating for a "great quantity" of Franklinia at the site. Jenkins listed eight reasons why he thought Franklinia an ideal planting for the park. It was named, he explained, for a "leading citizen" and held a "romantic history" associated with the explorations of the Bartrams. Furthermore, it would represent two states in two regions of the nation: Pennsylvania, the northern state where it was saved, and Georgia, the southern site of its demise. Beyond symbolic gestures, Jenkins noted it is free of pests, adaptable, flowers in fall, and is "beautiful and interesting in itself." But it was the way the species' name and history connect at this site, considered the birthplace of the nation, that was the true impetus for his suggestion. Franklinia, he stated, was associated with the first two names on the membership of the American Philosophical Society—Ben Franklin and John Bartram. To highlight this point, Jenkins' article was accompanied by a photograph of that society's October 1946 meeting (Figure 6.2). On the desk is a bouquet of Franklinia sitting just below Charles Willson Peale's portrait of Ben Franklin (Jenkins "Lost Franklin" 61).

Such powerful associations with people and places central to the creation of the nation have contributed to the longevity of garden Franklinia. Heritage is heard in its name and is the reason for its longevity. Jenkins, who would later become the president of the Historical Society of Pennsylvania, wove horticulture into historic preservation, connecting the plant to larger historic preservation efforts of his day. He imagined Franklinia an artifact in an outdoor museum dedicated to the nation's origins that would inspire the same associations as the surrounding colonial architecture. In 1951, botanist Alexander Anderson called Franklinia an "historic specimen." "This remarkable tree takes us back to the old colonial days of the eighteenth century," he wrote, "when George III was king and America little more than a vast unexplored continent..." (Anderson 81). From the late nineteenth into the twentieth century, historic preservation was a parallel movement to

Figure 6.2 Cuttings from the Franklin Tree beneath a portrait of its namesake.

environmental preservation, but here they are intertwined. Jenkins knew that encouraging more Franklinia plantings would help spread the species. This was not landscape-level, in situ, ecological preservation, but the careful propagation of a species to keep it from complete extinction.

Outside of the Altamaha, the most significant place associated with Franklinia is Bartram's home and gardens, on the outskirts of Philadelphia, where the single tree that is credited with saving the species grew. According to the Historic American Landscapes Survey, it is the "oldest surviving botanic garden" in the United States (John Bartram House and Garden HALS 4). John Bartram began the enterprise in 1728 with 102-acre farm that expanded as he became a collector who offered a wide variety of North American plants that he propagated and sold to European markets. The land, and the trade, stayed in the family through the early nineteenth century continuing and expanding as an educational center and a commercial nursery containing over 1,400 native plant species, ten greenhouses, and one thousand exotic species. In 1850, it was sold and kept as a private garden until its owner passed away in 1879. It was then that Thomas Meehan, who had worked at the garden, and botanists Charles S. Sargent campaigned to preserve it as a public park.

By the middle of the nineteenth century, Bartram's home and gardens were somewhat neglected. An 1882 illustration shows the home covered in vines and foliage, with grass growing tall and unkempt, in a rather romantic and scenic way. Though many arboreal "historic survivors" from Bartram's era remained, the Franklinia tree was damaged. In the 1880s, it seems a few cows had "horned the tree so seriously that it was all but dead" (Jenkins "Historical" 207). Meehan took the tree and propagated it by layering, an asexual form of reproduction in which a stem of a mother plant is wounded then pinned beneath the soil. The stem then begins to grow its own roots and can be separated from the mother. It was five years before Meehan established a new tree strong enough to be moved through this process. This intensive work meant Franklinia was not a nursery species easy to come by. "[T]o own a *Franklinia* tree during the Victorian era," Jenkins explained, "was to mark one as a horticultural aristocrat" (Jenkins "Historical" 207; Fry 112–113). Not only were the gardeners aristocrats, but the trees themselves held that sense of esteem. Jenkins wrote that all one hundred Franklinia growing around Philadelphia at the time of his writing (1933) could be traced to such "garden aristocrats" ("Historical" 207). Citing the failure of rediscovery searches to locate Franklinia in its Georgia habitat, Jenkins sets aside its loss in its natural landscape and suggests to make the best of the tree that remains, which he sees as inextricable from cultural heritage. "All in all it is a worthy monument and memorial of the 'First Civilized American' whom Philadelphia claims as its own," he wrote, referencing Benjamin Franklin. Viewing an acre of young Franklinia, he confided, "it bids fair to become an every day remembrance of the distinguished circle of Franklin and his botanical friends" ("Historical" 208).

Outside of its original habitat, Franklinia was caught up in a larger movement to revive and idealize the American colonial past. Scholars see the origins of this revival in the 1876 Centennial Exhibition, held in Philadelphia, to commemorate the anniversary of the signing of the Declaration of Independence. The fair spurred the travels of famous architects McKim, Mead, White, and Bigelow, who examined old buildings on an "antiquarian excursion" and spread a style dubbed Colonial Revival (Maynard 341–342). At the same time, people sought to preserve places of colonial history, most prominently the homes and birthplaces of founding fathers. The HALS report cited above couches preservation of Bartram's garden to this post-1876 "drive to establish an 'American canon' of places and heroes and to legitimize the country's place in the world order." When it opened as a public park, it was considered “a shrine of sorts to honor the contributions of the Bartrams” (43). In a 1913 article on the Bartrams’ home, the author describes vines creeping up the Doric columns, and includes images that couch the home in the picturesque mode, with a balance of natural elements surrounding its colonial architecture. “While the actual plants that Bartram cared for and loved are no longer in evidence,” the article stated, “the house has yet the appearance of being at peace with the plant world, of exhaling a faint suggestion that his spirit has not entirely passed away” (*Craftsman* 197). The editor of *The Craftsman*, Gustav Stickley, who was a leader of the American Arts and Crafts Movement, which preferred a simpler, pre-industrial aesthetic, chose to print this article directly in front of one that discussed the search for lost plant species, including Franklinia. This decision emphasizes the connections certain plants held to nostalgia and a certain memory of a simpler past.[6]

Preserving a place, like the Bartram’s estate, and preserving a tree in sporadic garden spaces are different acts of cultural preservation, just as reviving a species in situ is ecologically very different than propagating them for an arboretum. The tree at the Bartram home had its own sense of authenticity, and its narrative is clearly connected to historical events, akin to an artifact, but Franklinia trees in other garden contexts are distanced from this narrative in the same way a souvenir serves as a memento of another place and time. Most garden Franklinia are ripped from context, reminders of an imagined past. As Susan Stewart proposes, souvenirs are things associated with a narrative of which they are no longer a part. They derive from a place lost to time and an experience of that place that cannot be reproduced. Objects have the ability, she wrote, “to serve as traces of authentic experience… The souvenir distinguishes experiences. We do not need or desire souvenirs of events that are repeatable” (135). Like Franklinia, a souvenir is nothing without its narrative.

> The souvenir speaks to a context of origin through a language of longing, for it is not an object arising out of a need or use value; it is an object arising out of the necessarily insatiable demands of nostalgia.
>
> (135)

Trees as Memorials

During the U.S.’s Bicentennial year, 1976, families were encouraged to plant Franklinia as “living mementos” and to “ensure the preservation of this beautiful tree as part of the American heritage that future generations may enjoy” (*Ben Franklin*). Unlike those Jenkins proposed for Independence Hall, these Franklinia could be planted anywhere, by anyone, as an act that preserved both tree and memory of a certain heritage. One garden club that planted Franklinia that year nicknamed it the “Liberty Tree” (*Garden Club*). Franklinia plantings are connected to a larger movement to symbolically plant trees as "leafy monuments," which became an increasingly popular civic ritual in the early twentieth century (*George Washington).*

Even in the nineteenth century, trees were planted to remember individuals, such as the hackberry at Lincoln's house (1865) or the gingko at Grant's Tomb (1897). Important people also planted trees as markers of their visits. The Prince of Wales planted an English Elm in Central Park in 1920.[7] Often, trees were seen as appropriate ways to honor nature advocates; John Burroughs and Theodore Roosevelt have had trees planted in their names. Several clusters of trees planted around Philadelphia in the 1910s were to honor advancements in the animal rights movement.[8] Other groups organized to plant trees connected to important events or to honor trees that had witnessed historic moments. Historian Jared Farmer has written about such witness trees, emphasizing the aura of patriotism often placed upon them (Farmer "American Champion"). A single tree can be historic because it is old or, like Franklinia, because it is embedded in an historical narrative. Natural memorials to those who helped nature seem a fitting mode of commemoration, but trees were deployed as memorials after World War I not necessarily because of a connection to nature, but because they were seen as the right type of monument.

In 1922, Charles Lathrop Pack promoted trees as memorials, connecting the idea to historical precedent and democratic practices. "Tree planting to honor the heroic dead of the Great War, or others," he wrote in a publication for The American Tree Association (ATA), "has given the world a new form of monument—the memorial that lives" (Pack *Trees as Good* 108). Unlike monuments of stone or metal, he argued, "[t]he memorial tree is clothed in the finest of human sentiment…[f]or to-day and for generations yet unborn, the message of the memorial tree is the message of life" (108). Planting a tree as a "form of expression is possible to everyone," he wrote. It promised "community improvement as well as honor for soldier dead" (111). Another *American Forestry* article used similar language quoting Richard Lieber, Indiana's secretary of the Board of Forestry:

> Stone and bronze monuments may be heroic and military, they are more often vaingloriously dynastic in purpose. A monument of trees in a well-ordered grove is human and humane; it speaks the language of freemen. It is full of solace and hope to the bereaved. As a living breathing thing it speaks of victory over death.
>
> ("Trees for Memorials" 780)

All over the country, local organizations planted memorial trees in parks, lines of them along highways, and even forests, as efforts to remember the dead and as a way to beautify communities ("Trees for Memorials" 780). Children in Washington D.C. even planted a tree to remember the animals who died in the war, posing with the sapling and each holding a letter to spell "Be Kind to Animals" (*American Forestry* 26).

In 1936, Pack led the ATA in a campaign to plant trees for the sesquicentennial of the Constitution of the United States, receiving an Act of Congress to create a register. He compared planting trees to the founding fathers because both required looking “to the future and visualiz[ing] what will come in future centuries.” Pack argued that the most fitting way to mark this anniversary was by planting trees. “Speeches are of the moment,” he stated, “trees are of time.” (Pack *The Constitution* 18–19). George Washington’s birth bicentennial in 1932 was similarly celebrated and registered with the ATA honor roll. “Ten million monuments to a great man!” wrote Pack in characteristic prose,

> These are not to be monoliths of marble, not statues of static stone, or dun-colored copper and bronze, nor a eulogium written upon perishable parchment; but growing things, alive with a life emblematical of that living nation which the First President guided into being.

A flier advertising the effort encouraged this “patriotic service” and suggested specific trees that might be planted, noting Franklinia “is now available, after having been lost to horticulturalists for a century.” Franklinia’s inclusion in such patriotic plantings embeds it in the process of remembering and has spread it to new landscapes. The Georgia monument sounded the death knell too soon. The species is not dead or lost, but recirculated, migrating, and alive with meaning.

Laura McLauchlan argues in “A Multispecies Collective Planting Trees,” that conservation work to prevent extinction is often cast in “public narrative” as “heroic rescue”—an individual or a group fight to save a threatened species. Ursula Heise, too, talks of extinction narratives that are too simplistic and didactic. The story of Franklinia presented here is meant to expand the canon of extinction beyond tragedy to possibility. If extinct in the wild is not truly extinct, the terminology would benefit from a more critical conversation. Franklinia became a prized cultural object very early in American history—long before any activism to save species from extinction. It was never the topic of political, legal, or media attention; it lived quietly in gardens. While many made Bartram a hero in its salvation, it was not he alone that kept it alive. It was communal acts, as McLauchlan proposes, of “tending to life,” not a distance between nature and culture.

Franklinia has been as much a cultural object as a natural one. It was saved because of its cultural connections, not in the way bison were saved because they became a symbol of the frontier, but because individual people could plant, purchase, grow, and spread the species. It was embedded within the aesthetics of the garden and the nostalgia of the nation’s past. While McLauchlan’s essay is focused on larger ecologies, gardens similarly reflect “the work of tending to life, of refusing despair.” Franklinia continues to be a highly symbolic plant in twenty-first century gardens. Donald Goodman at Florida’s Kanapaha Botanical Garden says they grow Franklinia because:

"It's a beautiful plant. It's the story of classic Americana" (Faiello). To Johnathan Damery, editor of the *Arnoldia,* the journal of Harvard's Arnold Arboretum where Franklinia have bloomed since 1905, Franklinia reminds him of the dedication of staff botanists to document and describe plants for over a century, and the pivotal role the Bartrams played in keeping this species alive. "Imagine if the Bartrams hadn't seen this plant growing in the wild," he suggests, "imagine if they hadn't collected seed from it, to think that something like this could have been lost is quite devastating really" (Damery).

Listening to plant propagators, *preserved ex situ,* seems a better phrase than *extinct in the wild.* Both are accurate, but the latter suggests the very loss Damery is thankful did not occur. That phrase also removes wildness as an essence of what it means to be natural and suggests a more accurate relationship between humans and plants or, as Damery puts it, the "continuing living story of a relationship between people who observed and collected the plant and the plant as it continued to grow." Thus, couched in the heritage of horticulture and the history of human-plant interactions, Franklinia are artifacts. "I think one of the really cool things about plants like this," Damery explains, "is that unlike a painting in a museum...this Franklin tree, the object itself is living." Damery's goal is to go beyond plants as scientific and horticultural objects, to show them as "cultural objects...reviewing the ways in which the Arboretum, much like an art museum, is collecting these important stories of peoples, as well as the scientific work that goes along with them."

Franklinia's continued existence is celebrated by some, but to others its status is more elegiac. In the 1976 Oversight Hearing on the Endangered Species Act, Franklinia was chosen as "the symbol of the endangered plant program" because it was "extinct only in nature." (The report also used the phrase "extinct except in cultivation.") The replacement of "in the wild" for "only in nature" seems to distance the plant from the complex ideology of wilderness, but still invokes nature as a separate place from the garden. "Artificial cultivation" was not the answer, they argued, as it corrupted genetic diversity and should only be a "last resort." Habitat preservation was the priority, not ex situ species preservation. Referring to three Franklinia trees then growing in front of the Smithsonian Natural History Museum, the writers state: "So it continues to exist as a species—but rather like the almost extinct white tigers which are living out their lives in cages in the Washington zoo" ("Endangered Species").

Conservationists have tended to devalue the garden habitat in this manner. The places where Franklinia continue to grow are living natural landscapes and sites of public memory. Historian Seth Bruggeman talks about places of public history as landscapes that reinforce the "visitors' innate desires to see and touch the 'real' thing" (Bruggeman 201). As artifact, Franklinia is a real thing, a relic of the wilderness, a natural object held by historical figures. The power of the souvenir is that it internalizes the narrative; it is not just a thing, but a thing with a place-based story that gives it its meaning. It need not remain in its place of origin for those associations to flourish—in fact, souvenirs are made to be taken away. In teasing through the reasons Franklinia has remained an important garden specimen, its extinction story encourages examination of nature both near and far, domesticated and wild, in conversations about conservation. The Georgia memorial to the species and the reasons why it continued to grow in gardens as memorials both to themselves and to famous founders advocates for the power of narrative over naturalness in inspiring people toward preservation.

Notes

1 The work of William Cronon, Roderick Nash, and Max Oelschlager launched a vibrant field of inquiry into wilderness. Key texts include Cronon's "The Trouble with Wilderness"; Nash, *Wilderness and the American Mind*; Oelshlager, *The Idea of Wilderness*; and the edited volumes by Nelson and Caldicott. Roderick Nash, *Wilderness and the American Mind* (New Haven, CT: Yale University Press, 1967); on declension narratives, see Carolyn Merchant's "Reinventing Eden: Western Culture as Recovery Narrative," in William Cronon, ed. *Uncommon Ground.*

2 Ursula K. Heise. *Imagining Extinction: The Cultural Meanings of Endangered Species*; Genese Marie Sodikoff. *The Anthropology of Extinction: Essays on Culture and Species Death*; Dolly Jorgenson. *Rediscovering Lost Species in the Modern Age: Histories of Longing and Belonging*; Matthew Chrulew and Rick De Vos. eds. *Cultural Studies Review: Extinction*; Deborah Bird Rose, Thom van Dooren, and Matthew Chrulew. eds. *Extinction Studies: Stories of Time, Death, and Generations.*

3 Key ideas about monuments have been expressed by Christopher Woodward, Seth Bruggeman, and Denise Meringolo. On monuments to extinction, see Kelly Enright in *Cultural Studies Review* and Dolly Jorgenson.

4 Like Bartram, Franklin mostly worked in Philadelphia, some eight hundred miles away, and the intellectual center of the Revolutionary period. Franklin represented Georgia for a time, as well, making the connection all the more important as a regional expression of how they fit into the larger puzzle of American history.

5 Melissa Bailes contextualized Goldsmith's work; Christoph Irmscher's work dives deeply into the symbolisms inherent in colonial-era natural history, *The Poetics of Natural History.*

6 For scholarship on memory, see Michael Kamman's *Mystic Chords of Memory: The Transformation of Tradition in American Culture;* Seth Bruggeman's *Here, George Washington Was Born: Memory, Material Culture, and the Public History of National Monument.*

7 For more on specific trees, see Katherine Stanley Nicholson; Helena Pycior.

8 In 1916, a horse chestnut was planted in Philadelphia's Washington Square to honor the first Humane Week. The next year, three plantings honored women who worked with Henry Bergh in the Anti-Cruelty Movement. In 1918, a sycamore honored the American Anti-Vivisection Society and Animal Rescue League of Philadelphia (*American Forestry* 26, 336, 346).

Works Cited

American Forestry Association, *American Forestry,* Percival Sheldon Risdale, ed., vol. 26, 1920, pp. 336–346.

Anderson, Alexander. *The coming of the flowers.* Farrar, Straus and Young, New York, 1951.

Arbor Day Foundation. "A History of Arbor Day." Accessed 20 October 2022. https://www.arborday.org/celebrate/history.cfm

Architect of the Capitol. "Bartholdi Park and Fountain." Accessed 7 October 2022. https://www.aoc.gov/explore-capitol-campus/buildings-grounds/us-botanic-garden/bartholdi-park

"Arnold Arboretum." *Harvard University.* Accessed 20 October 2022. https://arboretum.harvard.edu/wp-content/uploads/2020/07/Franklin-Tree-transcript.pdf

Bailes, Melissa. "Literary Plagiarism and Scientific Orinality in the Trans-Atlantic Wilderness o Goldsmith, Aikin, and Barbauld." *Eighteenth-Century Studies*, vol. 49, Winter 2016, pp. 265–279.

Bennett, Thomas Peter. *Florida Explored: The Philadelphia Connection in Bartram's Tracks.* Mercer University Press, Macon, GA, 2019.

Botanic Gardens Conservation International. *State of the World's Trees*, September 2021.

Braund, Kathryn E. Holland. *The Ben Franklin Flowering Tree.* Unknown origin.

Braund, Kathryn E. Holland and Charlotte M. Porter, eds. *Fields of Vision: Essays on the Travels of William Bartram.* University of Alabama, Tuscaloosa, AL, 2010.

Bruggeman, Seth. *Here, George Washington Was Born: Memory, Material Culture, and the Public History of National Monument.* University of Georgia Press, Athens, GA, 2008.

Buell, Lawrence. *The Environmental Imagination: Thoreau, Nature Writing and the Formation of American Culture.* Belknap Press, Cambridge, MA, 1996.

Cronon, William. *Uncommon Ground: Rethinking the Human Place in Nature.* W.W. Norton & Co., New York, 1995.

Damery, Jonathon. "Frankling Tree." *Arnold Arboretum of Harvard University.* Accessed 27 October 2022. https://arboretum.harvard.edu/plant-bios/franklin-tree

Dunlap, Thomas. *Saving America's Wildlife.* Princeton University Press, Princeton, NJ, 1988.

"Endangered Species Oversight Hearing before the Subcommittee on Fisheries and Wildlife Conservation and the Environment of the Committee on Merchant Marine and Fisheries House of Representatives Ninety-Fourth Congress first session on Implementation and Administration of the Endangered Species Act and its amendments, and to review the problems and issues encountered." 1st–6th October 1975. Serial No. 94-17, U.S. Government Printing Office, 1976.

Enright, Kelly. "Exhibiting Extinction: Martha and the Monument, Two Modes of Remembering Nature." *Cultural Studies Review*, vol. 1, no. 25, 2019. pp. 154–171.

Entwisle, Tim. "Grow Franklinia, The Plant with No Home." *Garden Drum.* Accessed 2 July 2015. https://gardendrum.com/2015/07/02/grow-franklinia-the-plant-with-no-home

Evans, Dorothy. "A Rare Specimen Puts Down Roots—Ben Franklin's Tree Comes to Matthew Curtiss House." *The Newtown Bee.* Accessed 24 October 2003. https://www.newtownbee.com/10242003/a-rare-specimen-puts-down-roots-ben-franklins-tree-comes-to-matthew-curti/

Faiello, Mari. "Historic Tree in Bloom at Kanapaha Botanical Gardens." *The Gainesville Sun.* Accessed 17 July 2018. https://www.gainesville.com/news/20180717/historic-tree-in-bloom-at-kanapaha-botanical-gardens

Farmer, Jared. "American Champion Trees and the Wye Oak of Maryland." *Arcadia*, no. 3, Spring 2019. http://doi.org/10.5282/rcc/8502

Farmer, Jared. *Trees in Paradise: A California History*. W.W. Norton & Co., New York, 2013.

Fernald, Merritt, Lyndon. "Must All Rare Plants Suffer the Fate of Franklinia?" *Journal of the Franklin Institute*, vol. 226, no. 1353-28, September 1938, pp. 382–397.

Fry, Joel T. "Bartram's Tree: *Franklinia alatmaha*." *The Attention of a Traveller: Essays on William Bartram's 'Travels' and Legacy*, ed. Kathryn H. Braund. University of Alabama Press, 2022, pp. 73–113.Fry, Joel T. *Franklinia alatamaha, A History of That 'Very Curious' Shrub*. Bartram Broadside, Winter 2000.

Garden Club of Forest Hills Records, 1923–2011. "Guide to the Garden Club of Forest Hills Records, 1923–2011." AIS.2013.07, Historic Pittsburgh. Accessed 27 October 2022.

George Washington Bicentennial Tree Planting. The American Tree Association, 1932.

Goldsmith, Oliver. "The Deserted Village," *Poets of the English Language*. Viking Press, 1950. https://www.poetryfoundation.org/poems/44292/the-deserted-village

Harper, Francis and Arthur N. Leeds. "A Supplementary Chapter on Franklinia Alatamaha." *Bartonia*, vol. 19, 1937.

Heise, Ursula K. *Imagining Extinction: The Cultural Meanings of Endangered Species*. University of Chicago Press, Chicago, 2016.

Hoffmann, Nancy E. and John C. Van Horne, eds. *America's Curious Botanist; A Tercentennial Reappraisal of John Bartram*. American Philosophical Society, 2004.

Irmscher, Christoph. *The Poetics of Natural History*. Rutgers University Press, New Brunswick, NJ, 2019.Jenkins, Charles F. "The Historical Background of Franklin's Tree." *The Pennsylvania Magazine of History and Biography*, vol. LVII, no. 3, 1933, pp. 193–208.

Jenkins, Charles F. "The Lost Franklin Tree." *The Cornell Plantations*, vol. III, no. 4, Summer 1947, pp. 61–62.

"John Bartram House & Garden, 54th Street & Lindbergh Boulevard, Philadelphia, Philadelphia County, PA." Historic American Buildings Survey, Engineering Record, Landscapes Survey, Library of Congress, 2000. https://loc.gov/pictures/item/pa3904

Jorgenson, Dolly. *Rediscovering Lost Species in the Modern Age: Histories of Longing and Belonging*. MIT Press, Cambridge, MA, 2019.

Kammen, Michael. *Mystic Chords of Memory: The Transformation of Tradition in American Culture*. Vintage, New York, 1993.

Laboissière, Anna-Katharina. "Collect, Save, Adapt: Making and Unmaking Ex Situ Worlds." *Cultural Studies Review*, vol. 25, no. 1, 2019, pp. 65–84.

Lyell, Charles. *A Second Visit to the United States of North America*. Harper & Brothers, New York, 1849.

Magee, Judith. *The Art and Science of William Bartram*. Penn State University Press, University Park, PA, 2007.

Maynard, W. Barksdale. " 'Best, Lowliest Style!' The Early-Nineteenth-Century Rediscovery of American Colonial Architecture." *Journal of the Society of Architectural Historians*, vol. 59, no. 3, September 2000, pp. 338–357.

McLauchlan, Laura. "A Multispecies Collective Planting Trees: Tending to Life and Meaning Making Outside of the Conservation Heroic." *Cultural Studies Review*, vol. 25, no. 1, 2019, pp. 135–153.

Meringolo, Denise. *Museums, Monuments, and National Parks: Toward a New Geneaology of Public History.* University of Massachusetts Press, Amherst, MA, 2012.

Mighetto, Lisa. *Wild Animals and American Environmental Ethics.* University of Arizona Press, Tucson, AZ, 1991.

Nash, Roderick. *Wilderness and the American Mind.* Yale University Press, New Haven, CT, 1967.

Nelson, Michael P. and J. Baird Callicott, eds. *The Great New Wilderness Debate.* University of Georgia, Athens, GA, 1998.

Nelson, Michael P. and J. Baird Callicott, eds. *The Wilderness Debate Rages On.* University of Georgia Press, Athens, GA, 2008.

Nicholson, Katherine Stanley. *Historic American Trees.* Frye Publishing Company, Cambridge, MA, 1922.

Nixon, Richard. "Proclamation 4126—National Arbor Day," Online by Gerhard Peters and John T. Woolley, The American Presidency Project, 1976. https://www.presidency.ucsb.edu/documents/proclamation-4126-national-arbor-day

Oelschlaeger, Max. *The Idea of Wilderness: From Prehistory to the Age of Ecology.* Yale University Press, New Haven, CT, 1991.Pack, Charles Lathrop. *The Constitution of the United States with Tree Planting Instructions by the American Tree Association to mark the 1787–1937 Sesquicentennial.* American Tree Association, Washington DC, 1936, pp. 18–19.

Pack, Charles Lathrop. *Trees as Good Citizens.* The American Tree Association, 1922.

Parrish, Susan S. *American Curiosity: Cultures of Natural History in the Colonial British Atlantic World.* University of North Carolina Press, Durham, NC, 2006.

Peck, Robert McCracken. "Preserving Nature for Study and Display." *Stuffing Birds, Pressing Plants, Shaping Knowledge: Natural History in North America, 1730–1860. Transactions of the American Philosophical Society*, vol. 93, no. 4, 2003, pp, i–xviii, 1–113.

Pocock, Sarah. "New Report Reveals One Third of Tree Species Threatened with Extinction." *Global Trees Campaign.* Accessed 1 September 2021. https://globaltrees.org/news-blog/new-report-reveals-one-third-of-tree-species-threatened-with-extinction

Pycior, Helena. "The Making of the 'First Dog': President Warren G. Harding and Laddie Boy." *Animals and Society*, vol. 13, no. 2, pp. 108–138.

Rose, Deborah Bird, Thom van Dooren, and Matthew Chrulew, eds. *Extinction Studies: Stories of Time, Death, and Generations.* Columbia University Press, New York, 2017.

Sellars, Richard. *Preserving Nature in the National Parks: A History.* Yale University Press, New Haven, CT, 1997.

Slaughter, Thomas P. *The Natures of John and William Bartram.* University of Pennsylvania Press, University Park, PA, 1996.

Sodikoff, Genese Marie, ed. *The Anthropology of Extinction: Essays on Culture and Species Death.* Indiana University Press, Bloomington, IN, 2012.

Stewart, Susan. *On Longing: Narratives of the Miniature, the Gigantic, the Souvenir, the Collection.* Duke University Press, Durham, NC, 1992.

Stickley, Gustav. "A Picturesque Old House in Philadelphia Recalling the Adventurous Lives of John and William Bartram, Early American Botanist." *The Craftsman.* vol. 24, 1913, pp. 193–197.

Swanson, Drew A. "A Rhetoric of Ruin: Imagining and Reimagining the Georgia Coast." *Coastal Nature, Coastal Culture: Environmental Histories of the Georgia Coast*, eds. Paul S. Sutter and Paul M. Pressly. University of Georgia Press, Athens, GA, 2018, pp. 175–208.

"The Washington Bicentennial Tree Planting Project, 1732–1932." Biodiversity Library, W.G. NYCE, Pottstown, PA.

Thomson, Keith Stewart. " 'Marginalia': Benjamin Franklin's Lost Tree." *American Scientist*, vol. 78, May–June 1990, pp. 203–206.

Tonino, Leath. "We Only Protect What We Love: Michael Soule On The Vanishing Wilderness." *The Sun.* Accessed April 2018. Interview. https://www.thesunmagazine.org/issues/508/we-only-protect-what-we-love

Tredici, Peter Del. "Against All Odds: Growing *Franklinia* in Boston." *Arnoldia*, vol. 63, 2005. https://arboretum.harvard.edu/stories/against-all-odds-growing-franklinia-in-boston/

"Trees for Memorials." *American Forestry*, vol. 25, 1919, pp. 779–781.

Welch, Craig. "The Future of Forests." *National Geographic*, May 2022, pp. 72.

Whiteley, Peter M. "Epilogue: Prolegomenon for a New Totemism." *The Anthropology of Extinction: Essays on Culture and Species Death,* ed. Genese Marie Sodikoff. Indiana University Press, 2012, pp. 219–226.

Woodward, Christopher. *In Ruins: A Journey Through History, Art, and Literature.* Vintage, New York, 2003.

Wulf, Andrea. *The Brother Gardeners: A Generation of Gentlemen Naturalists and the Birth of An Obsession.* Vintage, New York, 2010.

7 Withnessing

Multispecies Approaches to Extinction, Testimony, and Bodies of Water

Toby Juliff

How might we think and feel through a shared experience of witnessing extinction? And how might we ethically entangle ourselves in witnessing the death of another, as they in turn provide testimony to the death of their witness? For whilst providing testimony for the lost has provoked much ethical deconstructive literatures (Levinas *Otherwise than Being or Beyond Essence*; Derrida *Demure*; Cixous *Veils*) there remains, I argue, an arena for "witnessing with" – as opposed to "witnessing for" – that might aide our understanding of the sixth great extinction event. In this event, the human shares their testimony *with and for* themselves *and* others. This shared witnessing, or "withnessing" in Nigerian Yoruba thinker Bayo Akomolafe's specific term, demands a radical re-orientation of human-centred testimony of extinction – in which the human mourns the loss of a specific, autonomous other – towards a shared act of mourning that conjures the long-extinct, to speak alongside the nearly extinct, to be heard by the not-yet extinct (Akomolafe *Slowing*).

In the work of French-Algerian theorist Jacques Derrida on his great friend the French novelist Maurice Blanchot, we see, or rather hear, what is at stake. Blanchot's 1999 short novella of testimony *The Instant of my Death* narrates an account in which, blindfolded and prepared for execution by firing squad, the author is unexpectedly spared by the arrival of partisan liberators. Blanchot's narrative speculates on this blindfolded witnessing of near-death, in which the accident of survival generates new speculative witnessing of actual deaths (those who were not spared). Blanchot's narrative raises the ethical question of how we might write into existence a death that is not our own, recounted by the nearly and soon to be dead. How might we provide witness to the death of another whilst alive? Derrida's counter-text response to Blanchot unpicks the ethical entanglement of literature and testimony by positing that:

> In essence, testimony is always autobiographical: it tells in the first person the sharable, and unsharable, secret of what happened to me, to me, to me alone, the absolute secret of what I was in a position to live, see, hear, touch, sense, and feel.
>
> (Derrida, *Demure* 29)

DOI: 10.4324/9781003315957-8

In this conversation is a range of haunting possibilities of new approaches to testimony, and to writing the ethical entanglement of witnessing the death of another.

Akomolafe's "withnessing" on the other hand finds resonance, if not analogy, in the works of others keen to radically challenge human-centred approaches to deconstructing human/more-than-human witnessing. Donna Haraway's consistent criticism of the work in earlier deconstruction of failing to see the more-than-human ["nonhuman" in Haraway's early pioneering work of human-animal studies] as more-than-*other* and failing to address shared experiences with the more-than-human as "companion" is important here too (*When Species Meet*). In developing new sets of relations, "withnessing" is an approach that drives new understanding of how humans mourn ecological death *alongside* more-than-human companions, a death that they are not only witness to but similarly subject to. And these mourning events might, to give deconstruction some redemptive dues, constitute a new "hauntology" (Derrida, *Specters*).

Within that ethical entanglement of witnessing, however, lies the problem that Akomolafe and Haraway – add to that theorist Karen Barad and others – have taken up. It is this "sharable and unsharable secret" that requires further unpicking in the stories of the sixth great extinction. Death is not a singular event but is, however, subject to the severing of specific relations that are shared and unshared across multiple instances of place and time. The insistence of the "I" in Derrida that the secret is to be shared and withheld by an individual might yet be resisted in favour of a "we" that testimonies of our own death are always co-authored with others. This co-authorship of testimony is, in the terms of this chapter, that "withnessing" which occurs when the human and more-than-human collectively mourn their specific and shared extinction.

This hauntology of new forms of more-than-human testimony, the conjuring of lost voices, and shared mourning has found great resonance in new creative, ecopoetical, and ecofeminist thought. This chapter seeks out new possibilities for thinking through withnessing the sixth extinction. In particular, our conjuring takes place in and in search of water. Through the work of ecofeminist thinker Astrida Neimanis, we will trace a phenomenology of bodies in a range of waters, arguing that through new forms of watery thinking we can sketch out a strategy for withnessing extinction alongside the more-than-human. Alongside Neimanis sits a range of new positions in Environmental and Blue Humanities that promote a cultural anthropology for extinction studies. Within these new natural and human-made bodies of waters, we'll seek gestures of solidarity and shared testimony to form this withnessing extinction that, through creative measures, might contribute new understanding of how we might learn to mourn together. In considering these ecopoetics of bodies in water, I argue that whilst many extinction narratives depend on the cultural and scientific context rooted in time and space (Sepkoski), hydrofeminist and hydrocommons reflections of "watery bodies" on land and in the ocean might enable new perspectives to emerge that call out to "unloved others" (Mitchell).

This chapter seeks to further explore the extinction poetics that exist within Neimanis' term "bodies of water". Within these discursive, figurative, and literal bodies, there are developments of what Blue Humanities scholar Steve Mentz will in turn call "saltwater aesthetics" and "Oceanic Humanities" that can be pushed inwards to land. These strategies of listening to and with the more-than-human might yet form a radical enchantment that constitutes a new ethical entanglement for working with "endling" species and the soon-to-be extinct in both land and water (Pyrry). As ecotheorist Dolly Jørgenson points out, such an endling – the last of its species – is "liminal, offering a viewpoint on the transition from life to death happening all around us, out of sight" (*Endling* 136). Endlings, in the terms we'll discuss here, make extinction concrete and render the space between the universal and personal porous and in flux.

We will seek this withnessing out through an examination of the work of lutruwitan/Tasmanian-based artist Selena de Carvahlo. In examining de Carvahlo's fieldwork in the shallow breeding grounds of the Giant Australian Cuttlefish [*Sepia apama*] and the in-land cemeteries of Campbelltown, Tasmania, I will argue that a common sensibility emerges in which a withnessing occurs that is shared and unshared with the soon-to-be extinct, the nearly-to-be extinct, and the eventually extinct. de Carvahlo's strategy of *listening with* the more-than-human stretches across oceanic and earthy terrain and sets out new gestures of solidarity through the generation of new creative works that place such shared exchanges within specific and mutual experiences. In employing new positions in ecofeminism, hydrofeminism, and cultural anthropology in the Environmental Humanities, this chapter argues that de Carvahlo's work exemplifies new possibilities for mourning and extinction. In conclusion, I argue that Neimanis' proposal for a "hydrocommons" might serve us just as well in-land as it might within the ocean and that de Carvahlo's practices of withnessing extend the terms of engagement with more-than-human witnesses to their own, and our shared and unshared extinctions to come.

Witnessing Immersion: Multispecies Testimonies of Water

For one week in June 2019, a group of fourteen artists set out to Point Lowly South Australia to meet and observe the gathering of *Sepia apama* (Giant Cuttlefish) (Figure 7.1). Set in the shallow waters of the Upper Spencer Gulf, as many as ten thousand cuttlefish congregate annually to breed amongst the temperate waters in plain sight, visible from both shore and water. Listed as "Near Threatened" in the IUCN Red List of endangered species, the annual cuttlefish migration has brought many divers, ecologists, ethologists, and now artists, to witness, record, observe, and reflect on the deep history of global migratory oceanic patterns of cuttlefish and other cephalopods.

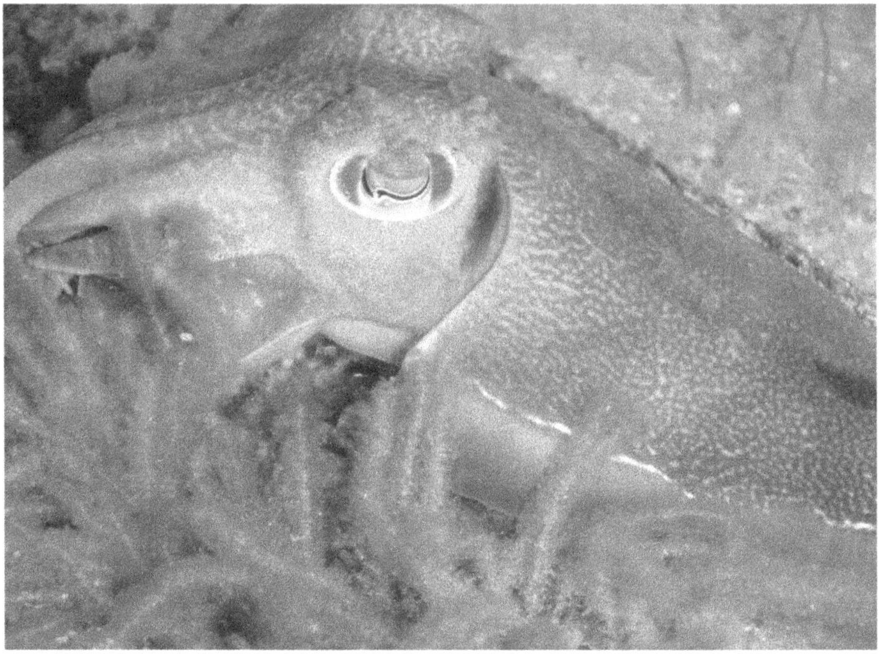

Figure 7.1 Sepia apama [Giant cuttlefish] at Rapid Bay Jetty, Gulf St Vincent, South Australia (2014). Author Peter Southwood. Used under Creative Commons Attribute Share-Alike 3.0.

Source: https://commons.wikimedia.org/wiki/File:Sepia_apama_P2023213.JPG. Accessed 1 November 2022.

In the biblical terms of Don Silcock, you can "literally [...] walk into the sea off the beach and the cephalopod version of Sodom and Gomorrah is all around you" (63–64).

Not far from Point Lowly is the town of Whyalla, home of a large Santos oil and gas refinery, a legacy BHP steelworks, and for more than 100 years, a large seaport. Still, a considerable exporter of ore, the port at Whyalla remains an active employer to the local town of Iron Knob. Constituting what Dolly Jørgenson will name "an energy landscape", Whyalla's industrial infrastructure has offered up its own unique aesthetic. Jørgenson reminds us that energy landscapes are spaces of experience, they are neither alien nor "out there" (et al. 10). And this is especially true in the case of Whyalla, where recent explosions of ecotourism intermingle with and are dependent on infrastructure (roads, hotels, food outlets, etc.) and service industries put in place to serve the energy landscape. In Jørgenson's terms, we disentangle at our peril:

> Like any other element of an infrastructure, aesthetics cannot be separated from its entanglement with the other elements that comprise

> an infrastructure—its politics, economy, materiality, its human and non-human inhabitants, and so on. Energy infrastructure is not randomly placed, or set in place to deliberately spite a landscape.
>
> (6)

Point Lowly sits on Barngarla Country, Aboriginal land that spans from Port Augusta 80 km to the East, and Port Lincoln 270 km to the Southwest. For many generations, the Barngarla custodians of the land on which Whyalla and its surrounds sit have witnessed the migration and breeding of Yaryardloo (cuttlefish); for more than fifty years, the Barngarla have continued to fight for sovereignty and the cessation of the dumping of nuclear waste on their ancient lands.

Since the Western scientific discovery of *Sepia apama* breeding grounds in the late 1990s, however, increasing ecotourism has bought new industries of year-round diving, boating, and walking. Its mild and dry winter climate draws thousands of tourists seeking respite from considerably wetter Victoria some four hundred kilometres to the East. Drawn by the profligate *Sepia apama*, the site has attracted the attention of many observers to the syzygy of late postindustrial extraction and cuttlefish breeding. Within this meeting of human and cuttlefish, a reciprocal witnessing occurs. As the cuttlefish observe each other, and we observe the cuttlefish and each other, we also share our witnessing of the destructive energy landscape that tests the resilience of *Sepia apama* and *homo sapiens* alike. For author Jennifer Mills, who has explored the phenomena of swimming alongside *Sepia apama* across multiple accounts, the experience of witnessing and sharing this breeding ground is both prurient and noticeable: "Our intrusion must have an effect. I have brought weights that will enable me to duck-dive easily, but it doesn't feel polite to descend to their level; they are busy with their cuttlefish business" (Mills *Swimming*).

Mills, across several accounts (*Swimming; Dyschronia)* of diving amongst the mating *Sepia apama*, notes the "600-million-year coincidence" of evolution that renders the eyes of the cuttlefish startlingly similar to our own. This shared act of witnessing each other with such familiarity brought about a bashful and human approach: "It [*Sepia apama*] seems to react just as a human's would: widening a little, studying the stranger for a moment, then looking politely away" (Mills, *Swimming*). The evolutionary design of the eye, which is no accident, provokes a shared experience of unsharable secrets. We – *homo sapiens* – look away at this act of the intimacy around us, as *Sepia apama* turns away from our gaze. If "bashful" is too much a speculative leap for this withnessing, it is noticeable for Mills that bodies in these shared waters appear to know why the other is there.

Developed out of Naarm/Melbourne-based sound research collective Liquid Architecture's collaboration with Lichen Kelp (Forum of Sensory Motion), a week-long residency called *cephaloPOD19* attracted fourteen artists to experience, explore, converse, and entangle themselves in the terranean and ocean surrounds of Whyalla. Joined by leading Aboriginal language specialist Ghil'ad Zuckermann, local Elders and members of the *Barngarla Language Advisory Committee*, and a rich mixture of interdisciplinary cephalopod scholars, the week-long residency and laboratory sought to explore *Sepia apama* through sonic experiment, documentary practice, more-than-human language, sensation, and wetsuit camaraderie.

Amongst the fourteen artists was Selena de Carvalho. As an interdisciplinary practitioner, de Carvahlo has explored the poetics of extinction across a series of site-based and installation practices that encounter new perspectives on more-than-human mourning of coming extinctions. With practices in deep inland and in shallowly ocean waters, there remains a vital sense of entanglement with watery bodies that blurs boundaries between advances in the Blue/Oceanic Humanities and orthodox positions within the Anthropocene, and between charismatic and unloved subjects of extinction.

Metaphoric Entanglements

As Thom van Dooren has noted, extinction is "a distinct unravelling of ways of life, a distinctive loss and set of changes and challenges that require situated and case-specific attention" (7). Whilst extinctions sever relations, a new kind of relationality emerges that extends beyond the act of violence that determined each extinction (De Vos). Further to this, van Dooren stresses that most extinctions appear as a "dull edge", not as a catastrophic singular event. Extinctions take time and take place, van Dooren argues, in a complex set of current and future extinctions that speak of "change and loss that occurs across multiple registers and in multiple forms both long before and well after this final death" (58). De Carvahlo's work extends that relationality into considerations of multiple registers and temporalities: whilst each extinction must be considered distinct and is entitled to its own distinctiveness, upcoming extinctions *live with the presence of the already extinct and the soon-to-be extinct.* No extinction is, in such terms, analogous with another even when they share the same time. But, as Barad will posit, deep entanglements occur nevertheless and demand continuous mourning (246). Such relations are, as Dolly Jørgenson has suggested, "messy" in which the living and nearly dead conjure the recently dead and long extinct (Jørgenson *Recovering*).

In the case of Whyalla, such metaphoric entanglements are palpable. As *Sepia apama* breed in decreasing numbers surrounded by industrial extraction infrastructures, so the "slow violence" (Nixon) of the sixth extinction is witnessed from off-shore with its *homo sapiens* kin. In contrast to the violent and rapid extinctions of the recent past, the adaptability and resilience of the cephalopodic and human body are subject to migratory patterns and camouflage that demand a slow and indistinct unravelling. The fossilised remains of the long extinct fuel the energy landscape that expulse toxic pollutants inhaled by the soon-to-be and later extinct. In this, there is no normalisation of violent extinction – rather a longer narrative that exposes multiple structures of harm and damage in which species that will become extinct sooner and later are active participants and witnesses to their own demise (Haraway *Staying*; Nixon).

Haraway's command to "make kin" develops strategies in resisting simplicity. They suggest that "it matters what matter we use to think other matters with" as a central tenet to these new imagined relations (*Staying* 12). Furthering this through a collaborative act of what they term "sympoiesis" – a process of making and *becoming-with* – Haraway's resists the "autopoiesis" of extinction or rather extinction ecologies that are contained, autonomous, and self-composing. Making kin, in these terms, is a resistance towards the sole authorship of extinction through a set of collaborative gestures that remind us that each extinction is not a severing of a single relation but might also be the bonding of a new kind of more-than-human relation.

Immersion

Kitted out in 3/2 mm wetsuits, hooded vests, and snorkels, the fourteen artists immersed themselves in the Spencer Gulf. As many of the artists deploy sound and visual practices, the lack of visibility and audibility made for a curiously quiet listening:

> It wasn't a complete silencing; floating in that other world, what we experienced was more of a muffling, a backgrounding, an underplaying of sound in its preferred medium of water. In that swaddled space, certain sounds were still recognisable, sometimes surprisingly so – friends' exclamations, the hiss of a passing scuba diver, the electro-static of crackling krill fields.
>
> (Zuvela and Lichen Kelp)

The act of immersion in water has attracted numerous quiet poetics. For Bonnie Tsui, however, underwater immersions mobilise "the awe and the terror, together" of the sublime (111). The high buoyancy of the wetsuits means mostly relatively shallow waters – a local South Australian colloquialism, "dodge tide" – can be explored. The loss of sensorial acuity and the continual pushing up the body for air, in such terms, becomes incredibly neutralising (Figure 7.2).

For Steve Mentz, such powerful immersions in water offer a unique perspective on precarity:

> Thinking from immersion helps us imagine and engage ecological instability. The surge, surrender, and velocity of wave-riding write a story in which tiny humans touch the hyperobject Ocean, and the two join, and swim, together. New narratives of ocean science remain to be written.
>
> (45)

Figure 7.2 Entering the water at Stoney Point (2019). Photo credit: Keelan O'Hehir.

Mentz, an advocate for wild swimming, "surf aesthetics", and Oceanic Humanities, reminds us that such immersions are never conducted alone, and always in solidarity. Of course, advocates of this approach to an imagined instability must also note that these immersions are only made possible through the intervention of technologies – wet and dry suits, oxygen and snorkels, masks and flippers – rendering such vulnerabilities heavily mediated. Our eyes, our skin, our lungs, and our bones are ill-suited to any great length of immersion in water, despite our evolutionary origins and its shared design in the eyes of *Sepia apama*.

On reflecting on their shared experience of the cuttlefish breeding and continued seismic exploration at Whyalla, the fourteen artists co-authored a reflection in which they noted:

> This ability to selectively deafen ourselves – even as a wetsuit side-effect – raises the inescapable politic of ocean listening. We sight-dependent land-creatures are unaccustomed to and generally uninterested in listening underwater – unless it's for military exercises or seismic exploration industries. These exceptions are fairly telling: these eavesdroppers silence others with explosives. There, in the water, one of our most important acts of listening took place, as a collective of temporary amphibians in the process of realising how imperfectly we were hearing, and chilled by more than the water temperature. We were in their space,

> the place where anthropogenic sound travels, swift and brutal. The noise our species makes with underwater exploration, weapons testing and shipping activity – that noise that kills, maims, strands, *pulps organs* and causes distressed cuttlefish to *change colour too quickly*. Self-deafened by our protective gear, we could still make out pale echoes of marine craft travelling through the Spencer Gulf to our location. These are sounds that our neighbouring non-human animals do not have the choice to not-hear, under-hear or un-hear. [...] What do we need to hear not just these sounds but also their contexts, the economic, political, military structures that make it so hard for us to unmute?
>
> (Zuvela and Lichen Kelp)

The human body is painfully inept at listening underwater. Recognising the sounds of *Sepia apama* congregating and mating amongst the sonic booms of continued exploration became difficult to distinguish for some. Increased marine traffic may have driven away many of the auditory-sensitive *Sepia apama* in the first few days of diving, the low-frequency high-intensity roar of engines providing even more distraction.

From Hydrofeminism to a Hydrocommons

Neimanis's posthuman feminist phenomenology reconceptualises "bodies of water". Neimanis reminds us that we are, after all, mostly water ourselves. And that, after all, water covers most of our planet. Neimanis develops a posthuman ethical framework that, in Steve Mentz's reading of this phenomenology, "insists on responsibility for vulnerable bodies while also recognizing that water's circulation and dissolution always exceeds bodily borders" (7). There is, therefore, something at stake for all of us in the entanglement and immersion politics of water. There is an embodied capacity to recognise the water in all of us, and still recognise that we are "still human" (Neimanis 26). We are human yes, but also the ocean, lake, and dam. We are alien bodies when in the water, we have neither the gills, lungs, nor auditory capacities to survive underwater for more than a few minutes at a time.

Neimanis traverses human and more-than-human frameworks of water, taking leads from the poststructuralist feminisms of Stacy Alaimo and Luce Irigaray, to propose an ethics of hydronic interaction that is "intimate" rather than "masterful" (112). Neimanis advances a position that stresses attentiveness to encountering more-than-human complexities through water. Intimate human scales can, in such terms, offer insight into global ecological disasters. This "antidote to Anthropocene water" refuses the "lithic phenomenon" (171) in which rock and inland water measures the impact of human activity more effectively than global tides. Neimanis deploys water through narrow non-Anthropocentric watery lenses: "In troubling the abstraction and homogenization of Anthropocene water, a posthuman feminist figuration – bodies of water – can suggest different orientations;

different selections of water's materialities and logics, and an amplification of those" (185).

New ethical frameworks such as Neimanis's hydrofeminism and Mentz's Blue/Oceanic Humanities resonate with the cephalopod orgy in the industrial waters of the Spencer Gulf. The energy landscape is inescapable and encroaches, minerally and acoustically, the breeding grounds of *Sepia apama*. Their more-than-human eye, as Mills reminded us above, is a 600-million-year coincidence of similarity to our own. They see the alien land, and the alien land-dwellers see them. It is a fluid "amniotic" of solidarity (Neimanis 29).

Anna Tsing's feminist anthropology of globalisation is significant here. Tsing argues that whilst the Anthropocene may be a global phenomenon, it is far from uniform in its expressions. From suburban gardens to urban floodings, and to oceanic plastic gyres, the Anthropocene may be far-reaching in its impact but is no less intimate or felt. In this, the Anthropocene is rendered as a granular expression of a set of complex ecological, gendered, and capital connections that result from the "combined but not fully synchronized effect of separate processes" (Tsing *Mapping*). In particular, Tsing's *Mushroom at the end of the World* considers an anthropology of capitalist destruction and collaborative survival that demands we look at the world-making possibilities generated by shared human and more-than-human experience. Advancing through a process Tsing names "assemblages", we might see in Whyalla a sense in which the resilient, camouflaged, and highly adaptable *Sepia apama* finds solidarity – not opposition – in the resilient, camouflaged, and highly adaptable *homo sapiens*. Together in the waters of the Upper Spencer Gulf, a multispecies assemblage emerges that disrupts and decentres the known environment. This disruption Tsing terms "unintentional design", an overlapping of different activities:

> Humans join others in making landscapes of unintentional design. As sites for more-than-human dramas, landscapes are radical tools for decentring human hubris. Landscapes are not backdrops for historical action: they are themselves active. Watching landscapes in formation shows humans joining other living beings in shaping worlds.
>
> (*Mushroom* 152)

Tsing's hopeful multispecies and granular approach to destructive capitalism sees the potential of finding solidarities between the more-than-human and human activities of disruption. When the *Sepia apama* breeding grounds were first formally identified in the late 90s, it was estimated that the nearby population sat somewhere at 170,000. By 2013, that population had dwindled to 13,500 (Mills *Swimming*). Since then, and thanks to the work of preservation working groups and temporary bans on commercial fishing, the population has returned to near that of the late 90s. At Whyalla, it remains uncertain if the mating grounds of the increasingly imperilled

Sepia apama or the proliferation of human eco-tourism will curtail or impede the expansion of the energy landscape, but Tsing's ethnography allows us to imagine for a second how together they might.

Cultural anthropologist Stefan Helmreich's argues that in an age of synthetic biology and rising sea levels (all enhanced by new technologies of listening), definitions of life, water, and sound demand a deconstruction. Helmreich identifies a series of conceptual "snapping points" that breach the boundaries of the empirical disciplines of marine science and the abstract social formations of culture, nature, and the digital and analogue experience of water. Cultural anthropological approaches such as these find an experience worthy of examination at Whyalla. The migratory patterns of *Sepia apama* collide with the migratory patterns of *homo sapiens*. Helmreich's expansive approach suggests that thinking through "waves" – wavelengths, social waves of new formations – allows us to ask essential question of scale and causation: "Waves, so queried, pose questions about structure versus agency" (*Wave Theory* 32). In the case of Whyalla, the "natural" waves are broken by the wakes and disturbances of ship movement and the vibrations of its energy landscape. The "natural" seismic energies of tectonic movement collide and are mistaken for the seismic fallout of continued mineral exploration and extraction. The scale of the migratory movement of *Sepia apama* is vast, and in its orgiastic mating intensely intimate.

New perspectives on Environmental Humanities continue to intersect with the "wet ontologies" of contemporary philosophy and the distinctively aquatic discipline of marine science (Mentz). Neimanis, Tsing, and Helmreich, each in their own disciplines, frame multispecies approaches to ethnographies of water that refuse the distinctiveness of "ocean" and "terrain". Though each may contribute new perspectives – Neimanis (feminist phenomenology), Tsing (Anthropocentric politics and global movement), Helmreich (anthropology, wave science) – there remain bonds in Darwin's "pleasant genealogy" that remind us that, "our ancestor was an animal which breathed water, had a swim bladder, a great swimming tail, an imperfect skull, and was undoubtedly a hermaphrodite" (453). And whilst the Blue in Blue Humanities remains marine and oceanic in its focus and privilege, we might yet still imagine that Darwinian evolution from water to land.

The entanglements of energy landscapes, cuttlefish mating, and human anthropologies make for an interesting week of anyone's time. In Whyalla, more than many places, these entanglements across language (human and more-than-human), sound (seismic, natural, manmade), and immersions find imperfectly new relations. And whilst *Sepia apama* appears, for now, more resilient and adaptable than was initially feared, the threat of extinction looms large. For the artists, contributing thinkers, custodians, and their allies, the laboratory conditions of onshore language acquisition, discussion and immersive dives, documents, and experiments become a site of multispecies inquiry that enables *homo sapiens* to reflect on their own imperilled extinction to come.

As the fourteen artists leave Whyalla after an exhausting week, each will take with them a set of approaches to examine these strange multispecies relationalities to the shore and develop new works. Some (David Haines and Joyce Hinterding) will explore sonic technologies of listening to and with multispecies agents, and others (Jannah Quill), the immersive synthesising of sounds that merge natural and manmade sonic activities. Some will explore the powerful resonances of Traditional Ecologic Knowledge (TEK) to reinstate Indigenous governance over sonic landscapes and ocean fronts through performance and storytelling (Libby Harward, Stéphanie Karbanyana Kanandekwe). Whyalla will resonate with works conducted on land and on sea, and in the case of Selena de Carvalho, in a graveyard 1,250 km away.

Part Two: Moving In-Land

Prasophyllum taphanyx (Figure 7.3) is an endemic species of wild orchid that grows in the heartlands of lutruwita/ Tasmania. With no more than three flowerings ever recorded in a single year – most years, just a single bloom – it is critically endangered. Its etymology, taphanyx is Ancient Greek: taphos meaning “grave” or “tomb” plus antyx meaning “edge” or “rim”. It is, in extinction terms, an “endling” species, the last of its kind. More specifically, and it will be important to be specific here, *Prasophyllum taphanyx* – commonly known as the “Graveside Leek orchid” – grows in the native grasslands of a single cemetery in Campbelltown, a small regional town (pop. 772) midway between the state capital nipaluna Hobart and the northern shoreline of the Bass Strait. It is, in terms of lutruwita/Tasmania anyway, as far from the ocean as it is possible to be.

For artist Selena de Carvalho, caring for the endling took place over a series of gestures that sought to generate a new understanding of the condition of shared witnessing and testimony that would result in an installation work entitled *Beware of Imposters (the secret life of flowers)* presented only a few months after the watery shallows of Whyalla. In the course of their caring, de Carvalho consulted with botanists, keepers of seed vaults (sadly, *Prasophyllum taphanyx* has proven thus far unsuitable for storing in global seed vaults), ethologists, and indigenous thinkers in generating a work that, in the artist’s own words, “hovers somewhere between installation, theatre and interactive art” (*Entangled* 22). Collecting stories, sharing water, imagining texts, and performing songs, the experience of the installation in a gallery some 100 km from Campbelltown takes the viewer into a darkened space decorated with a desk, chair, an old dictionary, and dried and discarded flowers sourced from the graveyard home. Strewn along the desk are pepper seeds (obtained from another graveyard in Naarm/Melbourne) and other miscellanies of mourning. Projected onto the desk are epigrams and poems generated at the graveyard and documentary images. Accompanying this timed experience is a soundtrack of a late 1980s song of longing

Figure 7.3 Prasophyllum taphanyx (graveside leek-orchid). Threatened Species Section (2022). Species Management Profile for Tasmania's Threatened Species Link. Department of Natural Resources and Environment Tasmania. Accessed 4 November 2022. https://www.threatenedspecieslink.tas.gov.au/Pages/Prasophyllum-taphanyx.aspx

(Chris Isaak's *Wicked Game*) that encourages a karaoke performance from the viewer/spectator.

In de Carvalho's own description, *Beware of Imposters* (Figure 7.4) constitutes a hauntology that seeks to "assemble a mixture of viewing, smelling, listening and improvising. Guided by the score, the audience is choreographed into becoming witnesses in a sympoetic assemblage of solidarity" (*Entangled* 35). Employing a mode of witnessing with, a withnessing, de Carvalho articulates the necessity of "response-ability" (Haraway *Staying* 28),

Figure 7.4 Selena de Carvalho. "Beware of Imposters" installation shot (2019). Photo provided by author with permission.

an approach to active witnessing the world-making potentiality and right of the orchid. de Carvalho themselves evokes the work of Akomolafe in advocating for this approach to an ethical encounter with the more-than-human other that encompasses narrative, immersion, and participation. Akomolafe's withnessing stresses the necessity of co-creating and co-designing encounters with the more-than-human. This commitment, which involves direct care (nutrients, physical protection) and indirect care (conversations), establishes a new regime for conjuring the more-than-human witness to the site of their own extinction.

Hydroacuity

Monica Gagliano is a researcher in evolutionary ecology at Southern Cross University (Australia) specialising in Biological Intelligence. Pioneering work in the fields of plant bioacoustics, more-than-human cognition, and Traditional Ecological Knowledge [TEK], Gagliano has argued persuasively that learning cannot be considered an exclusively animal condition. Through collaborative experiments with physicists, botanists, indigenous ecologists, and psychologists, Gagliano has proven that plants not only have the capacity to listen, but that they also have the capacity to learn from what they hear. They listen for water, they listen to each other, they listen to danger, and *they listen to us listening to them*. Talking to and *listening with* plants, Gagliano contends, is no longer in the domain of the speculative and mystical but is now leading the forefront of understanding cognition across human and more-than-human exchanges (Gagliano; Gagliano and Grimonprez and Depczynski and Renton).

Although far from the oceanic energy landscape of Whyalla, there is nonetheless a hydronic acuity at play here in the grassy graveyard of Campbelltown. As Gagliano has persuasively argued, plants not only emit distinctive voices, but that they are also capable of listening to acoustic landscapes of great complexity. In particular, plants have a unique capacity to listen for water through a complex network of roots that are able to respond to water (entreat and retreat) through listening, even when located far away from local water sources. Gagliano and other's cognitive plant ecology argues for the necessity to radicalise scientific engagement with the more-than-human, emphasising how the expanded notion of cognition might find reciprocal use in human/more-than-human exchange (Gagliano *Thus*).

Constructed only a few months after their experience at Whyalla, there is a peculiar shared and unshared immersion and listening for water that haunts *Beware of Imposters*. After adjusting to the low light of the installation room, a pair of headphones immerse the viewer in a soundscape of poetries written to the orchid. Part narrative, part speculation, part love letter to an absent other, these fragments are punctuated by projected images and documents:

> The ritual of the artwork attunes audiences through immersion within the temporarily assembled art-world. Engagement sensitises participants to the haunting qualities of loss. Yet, at the same time, experience nurtures knowledge and empathy through stories of human-to-human grief and those of more-than-human complexity and extinction. Conversation is embodied. Smells attract and repel. Perfumed bodies communicate chemical signals. Records of moments past hold voices, messages sent through mechanised memory are tuned and re-mixed, time is layered though symbolic notes and interpretation may be obscured. Still, care is delivered in the form of techno-ecological hauntings.
>
> (de Carvalho *Entangled* 30)

This chapter argues that though Whyalla and Campbelltown are far apart, and *Sepia apama* and *Prasophyllum taphanyx* are at different stages of their extinction, in de Carvalho they find a mutuality, a relational approach. By relational, we mean that knowing one another is a contingency made of social, sensory, embodied, and unexpected kinds. In this multispecies approach towards kinship, de Carvahlo seeks to animate a set of relations between human and more-than-human that attempts to decentre the human, whilst recognising the challenges of achieving that. In earlier work by Helmreich conducted with Kirksey, it could be said that Whyalla and

Campbelltown serve as vital "contact zones where lines separating nature from culture have broken down, where encounters between *homo sapiens* and other beings generate mutual ecologies and co-produced niches" (546). Each zone constitutes a generation of withnessing: in Whyalla, witnessing together in water; in Campbelltown, witnessing together for water.

Prasophyllum taphanyx is an endling, and *Beware of Imposters* is a love song to the last of its kind. This graveyard orchid may not survive another Tasmanian winter, but its rich capacity to listen for water, to seek it out through complex networks, is one that is shared by less-endangered, but still imperilled, kin. We know, thanks to Monica Gagliano and Selena de Carvalho, that such songs of longing might yet be heard.

Conclusions

Humans are terrible underwater listeners. And, we might argue, we're not all that good at it on land. And through deliberate or accidental evolutionary design, our eyes are remarkably similar to *Sepia apama*, yet we do not see quite so well underwater. But, for a moment perhaps, at Whyalla there is a shared withnessing that we can take with us as we emerge from the water and return to the land. Being with the nearly extinct [*Sepia apama*] and endling [orchid] both require a capacity of listening with; for water, in water, and under water.

Neimanis offers us a way through such distinctiveness of water and terrain. Amniotic relations afford us a new perspective on a hydrocommons that blur boundaries between "natural" and "man-made" bodies of water and the sites of negotiating the interbeing conditions shared by the human and the more-than-human. The critically endangered Campbelltown orchid *Prasophyllum taphanyx* shares water with the less than endangered, though similarly periled, *homo sapiens*. The act of shared listening for water, in de Carvalho's terms, is a "sympathetic being in the world with" (*Entangled* 6). Endemic extinctions are, we might argue, no more or less distinct – regardless of their terranean or oceanic contexts.

This chapter sought to argue for an approach to reading the extinction poetics of contemporary art that takes advantage of new perspectives on cultural anthropologies, hydrofeminism, and ethnographies of water. In placing ourselves in water, we might remind ourselves that we are already water. And no matter how inland we might find ourselves, we are always making kin with others that continue in this mutual search for water. In de Carvalho's practices at Whyalla and Campbelltown, there contains multiple exigencies: the immediate demand and long demise. If the Campbelltown orchid is the soon-to-be extinct, *Sepia apama* is the nearly extinct, *homo sapiens* the latter to be extinct. *Each one is urgent.* The demands remain bound by bodies of water. Extinction narratives are part of this shared and unsharable Anthropocene story. The mass extinction of our times represents more than death and irreversible loss – it is also bound up in the production, management, and monetisation of certain forms of life and favouring certain iconic species over "unloved others" (Mitchell 40). Darwin's powerful evolutionary axiom – that all terrestrial life was once aquatic – reminds us that, in Neimanis' terms, our hydrocommons might apply just as "pleasantly" in a graveyard far from the shore as it does in the oceanic mating grounds of a cuttlefish.

Works Cited

Akomolafe, Bayo. "Slowing Down in Urgent Times." *For the Wild Episode 155*, podcast, 22 February 2020.

Atkinson, Sally, et al. "Introducing with Microbes: From Witnessing to Withnessing." *SocArXiv*, 10 January 2021.

Barad, Karen. *Meeting the Universe Halfway: Quantum Physics and the Entanglement of Matter and Meaning*. Duke University Press, Durham, NC, 2007.

Barad, Karen. "No Small Matter: Mushroom Clouds, Ecologies of Nothingness, and Strange Topologies of Space Time Mattering." In Tsing A., Swanson H., Gan E., and Bubandt N. (eds), *Arts of Living on a Damaged Planet: Ghosts and Monsters of the Anthropocene*. University of Minnesota Press, Minneapolis, 2017, pp. 103–120.

Blanchot, Maurice. "The Instant of My Death." In Blanchot, Maurice and Derrida, Jacques (eds), *The Instant of My Death/Demure: Fiction and Testimony*. Stanford University Press, Redwood City, CA, 1999.

Cixous, Helene and Derrida, Jacques. *Veils*. Trans. Geoffrey Bennington. Stanford University Press, Redwood City, CA, 2002.

Darwin, Charles. "Letter to Charles Lyall, January 10, 1860." In Burkhardt, Frederick, et al. (ed). *Correspondence: The Correspondence of Charles Darwin*, 29 vol. Cambridge University Press, Cambridge, 2015.

de Carvalho Selena. "Beware of Imposters (the Secret Life of Flowers)." *Vegetal Entanglements II Antennae*, 52, 2021, pp. 45–57.

de Carvalho, Selena. *Entangled Solidarity: Poetic Gestures Embody Relational Methods of With-Nessing More-Than-Human Witnesses to the 6th Mass Extinction*, PhD Thesis. University of Tasmania, Australia, 2022.

Derrida, Jacques. "Demure: Fiction and Testimony." In Blanchot, Maurice and Derrida, Jacques (eds), *The Instant of My Death/Demure: Fiction and Testimony*. Stanford University Press, Redwood City, CA, 1999.

Derrida, Jacques. *Specters of Marx*. Chicago University Press, Chicago, IL, 1993.

De Vos, Rick. "Provocations from the Field – Extinction, Encountering and the Exigencies of Forgetting." *Animal Studies Journal*, 6(1), 2017, pp. 1–11.

Gagliano, Monica. *Thus Spoke the Plant*. North Atlantic Books, Berkeley, CA, 2018.

Gagliano, Monica, Grimonprez, Mavra, Depczynski, Martial and Renton, Michael. "Tuned In: Plant Roots Use Sound to Locate Water." *Oecologia*, 184, 2017, pp. 151–160.

Haraway, Donna. *Staying with the Trouble. Making Kin in the Chthulucene*. Duke University Press, 2016.

Haraway, Donna. *When Species Meet*. University of Minnesota Press, Minneapolis, 2008.

Helmreich, Stefan. *Alien Ocean: Anthropological Voyages in Microbial Seas*. University of California Press, Berkeley, CA, 2009.

Helmreich, Stefan. "The Genders of Waves." *WSQ: Women's Studies Quarterly*, 45(1–2), 2017, pp. 29–51.

Helmreich, Stefan. "Wave Theory ~ Social Theory." *Public Culture*, 32(2), 2020, pp. 287–325.

Jørgensen, Dolly. "Endling, the Power of the Last in an Extinction-Prone World." *Environmental Philosophy*, 14(1), Spring 2017, pp. 119–138.

Jørgensen, Dolly. *Recovering Lost Species in the Modern Age: Histories of Longing and Belonging*. MIT Press, Cambridge, MA, 2019.

Jørgensen, Dolly and Jørgensen, Finn Arne. "Aesthetics of Energy Landscapes." *Environment, Space, Place*, 10(1), 2018, pp. 1–14.

Kirksey, Eben and Helmreich, Stefan. "The Emergence of Multispecies Ethnography." *Cultural Anthropology*, 25, 2010, pp. 545–576.

Levinas, Emmanuel. *Otherwise than Being or Beyond Essence*. Trans. Lingis, Alphonso. Duquesne University Press, Pittsburgh, PA, 1974/1998.

Mentz, Steve. *Break up the Anthropocene*. University of Minnesota Press, Minneapolis, 2019.

Mills, Jennifer. *Dyschronia*. Picador, London, 2018.

Mills, Jennifer. "Swimming with Aliens." *Overland 130*. Overland Press, Melbourne, Autumn 2018. https://overland.org.au/previous-issues/issue-230/essay-jennifer-mills/

Mitchell, Audra. "Beyond Biodiversity and Species: Problematizing Extinction." *Theory, Culture & Society*, 33(5), 2016, pp. 23–42.

Neimanis, Astrida. *Bodies of Water: Posthuman Feminist Phenomenology*. Bloomsbury, London, 2017.

Pyyry, Noora and Raine, Aiava. "Enchantment as Fundamental Encounter: Wonder and the Radical Reordering of Subject/World." *Cultural Geographies*, 27(4), 2020, pp. 581–595.

Rose, Deborah Bird. "Shimmer: When All You Love Is Being Trashed." In Tsing A., Swanson H., Gan E., and Bubandt N. (eds), *Arts of Living on a Damaged Planet: Ghosts and Monsters of the Anthropocene*. University of Minnesota Press, Minneapolis, 2017, pp. 51–55.

Sepkoski, David. *Catastrophic Thinking: Extinction and the Value of Diversity from Darwin to the Anthropocene*. Chicago University Press, Chicago, IL, 2020.

Silcock, Don. "Giant Australian Cuttlefish." *X-Ray Magazine*, 82, November 2017, pp. 62–83.

The IUCN Red List Terms and Conditions of Use. Version 3, May 2017. Accessed 3 November 2022. https://www.iucnredlist.org/

Tsing, Anna Lowenhaupt. *Friction: An Ethnography of Global Connection*. Princeton University Press, Princeton, NJ, 2004.

Tsing, Anna Lowenhaupt. *The Mushroom at the End of the World: On the Possibility of Life in Capitalist Ruins*. Princeton University Press, Princeton, NJ, 2015.

Tsing, Anna Lowenhaupt, Deger, Jennifer, Keleman Saxena, Alder, and Zhou, Feifei. *Feral Atlas: The More-Than-Human Anthropocene*. Stanford University Press, Redwood City, CA, 2020.

Tsui, Bonnie. *Why we swim*. Rider, 2020.

van Dooren, Thom. *Flight Ways: Life and Loss at the Edge of Extinction*. Columbia University Press, New York, 2014.

Zuvela, Danni and Kelp, Lichen. *Field Notes from the Giant Cuttlefish Aggregation 2019: The First Unsaying*. Disclaimer. Liquid Architecture Press, Melbourne, June 2019.

8 Conjuring Up Ghost Species

On Photography and Extinction

William M. Adams, Shane McCorristine, and Adam Searle

Introduction: Spectres of Extinction

On 6 March 1938, a Cornell University doctoral student, James Tanner, took a series of photographs of a juvenile ivory-billed woodpecker (*Campephilus principalis*) sitting on the shoulder of his guide J. J. Kuhn in the Singer Tract in north-eastern Louisiana (Figure 8.1). They are the last known photographs of the bird, which was already on the brink of extinction. The last uncontested sighting was observed in 1944. While reports of the woodpecker continued into the current century, it is generally believed to have gone extinct soon afterwards, when the Singer Tract was logged (Bales; Tanner).

Figure 8.1 Ivory-billed woodpecker, photographed by James Tanner. Image public domain. Licensing information: https://www.fws.gov/media/166471

DOI: 10.4324/9781003315957-9

Tanner's images have become iconic records of the ivory-billed woodpecker, providing "generations of birders with an image laden with fragile, possibly doomed, hope" (Bales).

The possibility that the ivory-billed woodpecker has survived, somewhere, in the recovering swamp forests the American South has a considerable draw for the bird-tourism industry, focused on the Big Woods in Arkansas, where ornithologists reported a sighting in 2004 (Fitzpatrick 1,460). In 2009, a documentary film entitled *Ghost Bird* captured the intensity of birders' search for the missing species, and the woodpecker economy it supported. Gandy finds this trope of the ivory-billed woodpecker as a "ghost bird" problematic, pointing out the way in which such language shrouds the racialised history of the landscape of the Mississippi Delta "in a kind of ornithological haunting driven by the intense desire of both amateur enthusiasts and professional scientists to catch a glimpse of the bird" (372). This mode of haunting, in Gandy's understanding, is riven with the chimera of "European adventurism" (373).

Current rates of extinction are far higher than the pre-human background rate, making our era of extinction very different from previous planetary catastrophes. The idea that, in the Anthropocene epoch, we are haunted by extinct species has become a prevalent rhetorical frame for discussions of the problem of biodiversity loss (Barrow; Heise; McCorristine and Adams; Searle "Hunting Ghosts"; "Spectral Ecologies"). Living through the "sixth mass extinction"—a period defined by ecological destruction, loss and absence (Searle "Absence")—the figure of the ghost has emerged as a fruitful analytical lens to consider how past, present, and future ecologies shape each other (Tsing et al. eds.).

The boundary between presence and absence, and the significance of its policing, explored extensively by cultural geographers (e.g., Wylie; McCormack), is useful for examining the cultural dimensions of extinction. For instance, Ginn, Beisel, and Barua (117–118) argue that the modern era of extinction creates opportunities for scholars to imagine and explore spaces of absence and disappearance, and van Dooren sees extinction as a kind of haunting, "a slow unravelling of intimately entangled ways of life that began long before the death of the last individual and continues to ripple forward long afterward" (12). Vinciane Despret suggests that "the world dies from each absence" ("Afterword" 219)—extinctions do not signal "the end", but rather, something entirely novel, which is haunted by ecologies of the past and future.

Conservation stories draw extensively on notions of the spectral (e.g., the Fairyland Trust, a registered UK charity that aims to reconnect people with nature through a sense of enchantment). These attempts to see species that might or might not be extinct are engagements with liminal and spectral beings in the realm of the imagination (Heise 19–54). Rare and extinct species, it is argued, dwell as spectres on a border between the visible and the invisible (Yusoff 585). Charismatic extinct species such as the dodo are the darlings of conservation headline writers, who frequently claim another species is "going the way of the dodo" (Guasco 1,056). These species seemingly destined for extinction provide the narrative counterpoint to classic redemptive stories of "rescues" (see Jørgensen *Recovering*), for example, the American bison, shot to the edge of extinction and then captive-bred and reintroduced across North America.

Yet, while species such as the dodo are known definitively to be extinct, the survival of many other species remains equivocal. The continued survival (or extinction) of species that are uncommon, hard to observe or to identify, is difficult to establish (see Bastian). Such species enter a limbo state that might be described as "missing, feared extinct". Reports of their absence are open-ended, the result of surveys that may or may not be judged complete enough. Such species, whose extinction is increasingly widely believed, exert a powerful hold on conservationists because they have an enduring spectral presence in landscapes, made present through storytelling and the search for evidence of survival (McCorristine and Adams 102–104). Their disappearance tends to produce ghostly effects and memorials: traces, signs, and clues are left behind, provoking reports of reappearance, while at the same time, artists mourn and commemorate their absence.

Gandy observes that the idea of the ivory-billed woodpecker as a "ghostly presence" in the flooded forest represents "a form of ecological nostalgia for an imaginary web of life that predates the imposition of European modernity" (382). Nostalgia is a powerful force in memorial cultures of extinction, eliciting emotional responses that can occlude and reveal our violence against nature in the past. One of the chief currencies of nostalgia in the era of extinction is the photograph. Photography has a particular and important place in cultural responses to the disappearance of non-human animals because of its capacity to evoke lost worlds and lives. Reinert, for example, examines the links between haunting and photography through a widely circulated image of a lesser white-fronted goose (*Anser erythropus*) named "Imre" who was poached in Russia in 2006. In similar ways, Tanner's photographs of the individual ivory-billed woodpecker in 1938 have become intertwined, in knotty and emotional ways, with a species-level story of extinction and habitat destruction. As evidenced in the deluge of Anthropocene exhibitions, artworks, and mourning rituals in the past twenty years (see Ibbotson; Sklair), memorialisation leans heavily on photographic lens to scale down the enormity of extinction to an indexical representation (see Kember; Mudie).

In this chapter, we explore the role that photography plays in conjuring up the ghosts of extinct and near-extinct species, enabling them to have powerful afterlives in conservation communications and public memory. There are recurring debates and claims surrounding the physical presence of the species we mention, but we are interested in what contexts these after-living "ghost species" are sensed or generated by photographs. We proceed by reflecting broadly on the particular characteristics of photography that enable it to play this role, focusing on how the technology makes determinations of presence and absence, before turning to the documentation of "endlings". In the concluding section, we discuss examples of how photography features in recent artistic responses to extinction and biodiversity loss in the Anthropocene.

Photography, Extinction, and Haunting

Conservation fears about extinction date back to the start of the modern conservation movement in the nineteenth century (see Adams *Against*; Barrow). These fears arose at the same time as photographic processes became widespread and conservationists shared with practitioners of the new technology a concern to archive, preserve, and curate that which was classed as extinct or at risk. Indeed, Carleton Watkins's early landscape photography in 1860s California was fundamental in garnishing widespread support for the protection of Yosemite's "wilderness", just as contemporary conservation movements utilise photography to communicate the ecological condition in a plethora of forms which often subvert the tropes of a sublime or pristine nature (DeLuca and Demo 242). The concern to capture the vulnerable traces of deep time in order to conserve aspects for future generations can be seen in early photographs of antiquities and ruins in the Middle East (Lyons et al. eds. 28) and daguerreotypes of shells and fossils (Zylinska 56).

Until the end of the nineteenth century, public perception of the diversity of animal life depended on physical specimens, preserved through taxidermy in museums and alive in zoological collections (zoos and aquaria). Concern at the depletion of "game" and the extinction of species led to a concern for *in situ* conservation, in national parks and nature reserves (Adams *Against*). Over the first half of the twentieth century, a gradual shift took place from the collection of physical specimens to the use of cameras to capture, preserve, and present wild animals. Although big game hunters continued to hunt rare animals and take trophy photographs of their kills—once described as "hunting with a camera" (J. R. Ryan 206)—others began to photograph charismatic species in the wild, rather than kill them. In Africa, the safari industry evolved to focus primarily on game viewing and not hunting tourism. Wildlife photography became a mainstay of natural history publishing and wildlife documentary film a recognised genre both of cinema and of television film making (see Mitman *Reel Nature*; Attenborough). This shift in meaning echoes the point Sontag made about the history of photography: "When we are afraid, we shoot. But when we are nostalgic, we take pictures" (11).

As photography gained ground and became the primary means of documenting and representing wild animals to the public, photographs of declining, "threatened", or endangered species became more important to conservation. Photographs of rare species became tools for conservationists, capable of inspiring action and nostalgia in equal measure.

For instance, Carl Akeley, the taxidermist famous for his dioramas in the African Hall of the American Museum of Natural History in New York, also knew the power of photography (Haraway 321–325). He wrote: "what man seems bent upon destroying with his gun can at least be rescued from complete oblivion and given the illusion of reality through the camera" (qtd. in Dunaway 227). For Akeley, photographs of rare animals like the mountain gorilla were powerful records of animals facing extinction. Akeley's work exemplifies the close linkages between photography and taxidermy (Behr 51), and the way photography became an invaluable documentary tool, influencing the development of taxidermic practice (see Alvey).

Photography changed the forms taken by both the determination of presence or absence and the determination of memorialisation, with a transition from the preserved skins or bodies of taxidermy to the two-dimensional images of photography. Species that became extinct before the invention of photography are still only present to scientists through specimens, skins, descriptions, or drawings. Even today they can only be evoked through imaginative representations and framings. Many species that became extinct in the age of the camera, on the other hand, are known by their photographs, and this places them in a different order of being, with individuals part of a regime of endless mechanical reproduction. Furthermore, because this genre of photography has been used as a form of portraiture in conservation history, images of individual members of extinct species have been alternately named, storied, designated "the last" of their kind (see Pyne; Jørgensen "Endling"), or otherwise given afterlives by virtue of their photographic capture.

So how does photography enable extinct species to be conjured from presence to absence (and from absence back to presence again)? Photographs provide not only a spectral view of absence but also a truth-claim about presence. Photographic evidence is central to efforts to establish the continued presence of scarce or cryptic species. A good example of this can be seen in automated fixed digital cameras (trail cameras or camera traps), which have become increasingly widely used in conservation surveys to record the presence of nocturnal, elusive, remotely located, or critically endangered species (Adams "Geographies"). Thus, in Cambodia, a small mammal, the yellow-bellied weasel (*Mustela kathiah*), was photographed in 2014 after eight thousand hours of camera trap recordings (Ludtke). The key advantage of these camera traps is that once installed they operate automatically, and only demand human presence to change batteries and download data. Remote online identification of the presence or identity of species by members of the public comprises new practices of citizen science, and is allowing the development of algorithms to identify species individuals automatically (Rowcliffe and Carbone 185–186).

Through camera traps, the human watcher is both absent and present. In fact, the implication is that it is only by humans being absent, or spectrally present through the technology and remote observation, that the animal will appear at all. Camera traps in the wild are therefore profoundly uncertain things—they register presence if bodies present themselves, but a lack of photographs does not prove absence. Indeed, animals sometimes suspect the presence of humans through trace smells left on the cameras and the sounds they make, enabling them to avoid its gaze, thereby remaining uncaptured and unrepresented (Meek et al.; Wegge et al.). The spectrality of these species demonstrates how photography in the age of extinction has a dual role in certifying presence and life, while also acting as an archive of the dead.

This dual role can be seen in historic photographs of extinct species, especially those thought to be "the last" of their kind, which tend to have considerable cultural power. Many images of extinct species have become iconic, featuring in documentaries, conservation campaigns, and creative representations of extinction. In Fuller's *Lost Animals: Extinction and the Photographic Record*, a sepia-tinted photograph taken in New Zealand around 1909 shows a Laughing Owl (*Ninox albifacies*) crouched in a rocky crevice, grasping a dead mouse in its beak: this individual bird was the only member of the species ever photographed in the wild. A few years later, the last known living bird was collected as a museum specimen. The 1909 photographic image therefore stares out across an absolute divide and seems to challenge it—life meeting death. In 2014, the New Zealand-based artist Sally Ann McIntyre used this photograph as a point of departure for her memorial to the species. After visiting a mounted specimen of the species in a museum, McIntyre recorded "one minute radio silence" for the laughing owl on site, using the paranormal investigation method of "Electronic Voice Phenomena". This would, McIntyre suggested, "[give] space to its absence" and mark the centenary of the bird's extinction by "listening in to the one-hundred-year lack of any signal between 1914 and 2014".

Alongside artistic responses, the genre of archival rediscovery is a powerful means of communicating the afterlives of extinct species. Take the silent, black-and-white film recording from 1918 of an individual heath hen (*Tympanuchus cupido cupido*). Filmed by the Massachusetts Department of Conservation, this was part of a doomed attempt to study and preserve a species that had once roamed New Hampshire and Virginia: it became extinct in 1932, and the photography was archived and lost to sight. In 2014, the footage was rediscovered and digitised, thrilling conservationists who saw a vanished bird suddenly "come to life" (Johnson). In a similar story, still images extracted from moving footage of an imperial woodpecker (*Campephilus imperialis*) taken in Louisiana in 1956 were only rediscovered and publicised widely after 1997 (Fuller 113). This footage has been described as possibly "our last chance to gain substantial knowledge of the species" (Lammertink et al. 675). Having a status that lies somewhere between life and death, these photographs and reels of gone species suggest that the power underlying extinction images is "apparitional"—that is, they portray an absence that is not actually absolute and because of this they draw us into emotional relationships with the subjects.

The continued cultural prominence of absent biota unsettles firm distinctions between presence/absence and past/present; as such, extinction is seen as a "non-linear and paradoxical process in that it leaves traces, signs and clues, and can provoke recurring reappearance" (McCorristine and Adams 105). Reappearances, documented by photographs, shape the experience of conservation communications in such a way that "geographies after extinction are haunted geographies" (Garlick and Symons 312). Revenant species in photographs, reels, or sightings are never the same as those which disappeared historically: they emerge in distinct cultural and ecological contexts (Searle "Spectral" 170–171) and produce distinct affects and relations with those left behind to mourn and remember (Garlick 236). Take the example of the Eastern cougar (*Puma concolor couguar*), which was last verifiably recorded in 1938 and officially declared extinct by the US Fish and Wildlife Service in 2015. Cougar and "big cat" networks across the USA continue to share photographs of suspected cougars, especially through dedicated Facebook communities (e.g., "Pennsylvania Cougar Sightings"). Although the Department of Environmental Conservation in New York State warns against believing hoax/false images, at the same time officials request people to submit any photographs of suspected cougars they have.

Photographs of extinct species belong to another order of presence from museum specimens, or drawings, one that accords with our modern experience of loss as mediated by technology. In her memorial to the laughing owl, McIntyre recognised that photographs of vanished species have the power to grab our attention in ways that stuffed specimens, engravings, or other means of representation do not possess. This is because, firstly, the technology of photography has potent historical and philosophical links with death and resurrection, and, secondly, because photographs of extinct species demonstrate a continuing spectral, or life-giving presence for conservationists.

Photography has strong links with death and haunting because of the apparently magical way that it captures the past (Roberts 393–394). A photograph is a trace of something that has been, and that then made a photochemical impression on a sensitive plate. Photographic content has dual status as a frozen snapshot of a reality that has passed, and a hyper-mobile, even lively, re-presentation that reaches into the future. In creating disembodied images of bodies that were once alive and breathing, photography conjures its subjects into life, even as it represents death: through a photograph, an absence becomes a presence—this is a truth amply demonstrated by two popular spectral practices in nineteenth-century Britain and North America—"spirit photography" (in which the dead return with life) and "postmortem photography" (in which the dead remain with life).

Photographs of this kind have a liminal status as a "living image of a dead thing" (Barthes 79). However, despite their interest in haunting and in/visibility, for a long time they paid little attention to the appearance/disappearance of animals. Elsewhere we have addressed this shortcoming, elaborating links between photography, conservation, and the extinction of animals (McCorristine and Adams; Searle "Spectral"). Other scholars have explored the links between the development of photography and the idea of animals becoming spectral. In his essay "Why look at animals?" Berger traces the way that, under modern capitalism, most animal species vanish from everyday life only to reappear as images and spectacles in zoos, where they "constitute the living monument to their own disappearance" (26). Building on Berger's argument, Lippit asserts that animals now exist in a state of "*perpetual vanishing*" (1), lingering on as ghosts in cinema and photography. In the age of photography and extinction, she suggests, animals have merged with their means of technological reproduction: media technologies now serve as "virtual shelters for displaced animals. In this manner, technology and ultimately the cinema came to determine a vast mausoleum for animal being" (187).

The designation of extinction adds another layer of meaning to photographs of vanishing animals. In a technical sense, all photographs are spectral because they expose the absence of bodies and register the time between capture and development (McCormack 650). Barthes goes further and suggests that photography turns its subjects into ghosts because they bring the dead back to life. Photographs of animals that became extinct, then, are haunting because they capture the lively trace of a species now gone—they show it existing in the era of photography. Fuller's book gathers together some of these haunting images. They include a blurry photograph (the only one in existence) of the Alaotra grebe (*Tachybaptus rufolavatus*) in 1985 and a forlorn-looking Quagga (*Equus quagga quagga*) in London Zoo in 1870. These photographs have an unsettling permanence—they are, as Barthes puts it, "certificate[s] of presence" (87). They create a sense of catastrophe and loss that is palpable and more powerful than the feelings generated from an engraving or narrative about an extinct species.

This is because, in photographs, the dead immediately and repeatedly return: this is why they have a talismanic role in how we mourn and remember. Indeed, the tangible role of photographs in memorial cultures is recognised by authoritarian regimes that "ghost" citizens by, for instance, confiscating the photographs of people that have been abducted by agents of the State (for examples from Argentina, see Gordon 103). Although in some cases photographs were casually taken of species that were not expected to suddenly disappear, many photographers and conservationists sought out species that were on the brink, a status usually indicated by their presence in zoos like the several thylacines (*Thylacinus cynocephalus*) photographed in Beaumaris Zoo, Tasmania. This de-wilding of "the last" brings added poignancy to the scenes and adds to the sense of portraiture around the photographs: before the extinction of species, individuals seem to pose for spectral photographs. They reach into a future-time beyond their death. Photographs of extinct animals have emotional impacts on viewers, but to judge by their regular appearance in conservation communications, photographs of the last surviving individuals of species—so-called "endlings" (Jørgensen "Endling"; Pyne)—have significant power in conservation discourse. These photographs make extinction iconic and draw us into a relationship with the subject captured. This is clearly the case with the thylacine, a species that, as Freeman shows in *Paper Tiger*, became extinct despite a diverse media of drawings, engravings, photographs, and film footage. Through this case, Freeman explicates how visual cultures profoundly shape conservation action and human engagements with endangered species. These representations shape memorial cultures around the thylacine: for instance, photographs and film reels of the Zoo thylacines feature prominently in displays in the South Australian Museum, the National Museum of Australia, and the Tasmanian Museum and Art Gallery.

Perhaps the most famous endling is a captive passenger pigeon named Martha. Passenger pigeons (*Ectopistes migratorius*) once crossed America in flocks that numbered billions of individuals, but hunting and habitat destruction by humans caused a sudden collapse in bird numbers in the late nineteenth century. By 1910, the last surviving member of a group held at the Cincinnati Zoo. She was given a name, Martha, and became a celebrity for visitors and was photographed by Enno Meyer in 1912. Martha's life ended on 1 September 1914, when she was aged around 29. This photograph shows Martha alone in captivity—this was the intention of the photographer. The poignancy in the photograph (a factor beyond the intention of the photographer, which Barthes calls *punctum*) came later, with the realisation that she was "the last". Martha died two years after the photograph was taken, and, as Cokinos relates in *Hope is the Thing with Feathers*, with her individual death the world lost an irretrievable evolutionary heritage.

Photographs of people who have died unsettle the relationship between absence and presence: here they are, alive and well, in images that trace their presence. In this sense, people become ghosts in photographs, haunting the present from the past. However, a photograph of the last member of an animal species is different. This is because, on the one hand, the subject in the photograph stands for the whole species, and on the other hand, it is a bodily presence with a storied life, not just an abstract idea. Can listening to the stories of the passenger pigeon in conjunction with its conjured presence disrupt our inherent speciesism? This might enable us to mourn their loss as something that impoverishes a shared ecological community. Extinction would then be made "real, imminent, sensible and a matter of survival for all life" (J. C. Ryan 76).

Photographs of endlings are therefore magical images for conservationists—at once signs of individual life and icons of species loss. They are presented as both "real and ideal in an enunciative present, one which is separated from both the past, a time of presence, and the future, a time of absence" (De Vos 190). Meyer's photograph of Martha is a scene of conjuration: she will always be the last of her kind—the catastrophe of the extinction of the passenger pigeon has already happened. Yet, she also becomes a ghost "whose expected return repeats itself, again and again" (Derrida 10). When found dead in her cage, Martha's body was shipped in a block of ice to the Smithsonian Museum in Washington DC, where she was preserved as a taxidermy specimen. A rotating 360-degree image of the mounted body is presented on the museum's website, as part of the museum's *Objects of Wonder* exhibit.[1] Through techniques and technologies of mass culture in the Internet age, the digital images of a preserved specimen conjure new connections with Martha as a dead object of scientific intrigue. As noted by Ibbotson, these "artefacts of natural histories" are themselves co-constituted with "visual clues" that shape memorial cultures (85).

Every reproduction of the photograph conjures up her spectral presence. Photographs of endlings like Martha or "Benjamin", the thylacine (d. 1936), unsettle the notion that these species are utterly gone. This is because photography enshrines them in a "virtual shelter" (Lippit 187) where they live on and become endlessly reproduced as an object of memorial culture—its traces are omnipresent in Tasmanian culture from its presence on the state's emblem to bottles of Cascade beer. Totemically represented by named and photographed individuals, it is no coincidence that these are three of the top extinct species earmarked for biological resurrection through techniques associated with de-extinction (Brown; Zimmer). The contemporary appeal of the thylacine as an extinct emblem of Tasmanian past ecology and synthesised speculative futures is in part driven by an overtly visual emphasis in memorial culture, as "in de-extinction discourses, images bear the representational burden of what cannot be seen" (Ibbotson 84). For Ibbotson, these images occupy an "extinct uncanny" where visual representations "destabilise previous certainties such as the finality of death and extinction" (ibid. 90).

Art, Photography, and Extinction Afterlives

In this final section, we explore how affective atmospheres of loss and nostalgia can be evoked in extinction photographs by the particular methods and media used. Photographs of endangered species produced for art and science seek to conjure presences in the wake of their biological absence, yet the nature of these relations differ in their aesthetic, narrative, and intended publics. One key example is the artistic use of sepia-tinted or "Victorian" filters and connections to the daguerreotype—one of the earliest incarnations of the photographic process (Bezan 437). Visual artist Christina Seely's played with these affects in her exhibition *Next of Kin: Seeing Extinction through the Artist's Lens* at the Museum of Natural

History, Harvard University. Seely's daguerreotype portraits were of endangered species, a process chosen because the images could be rendered on a mirrored surface so that the image can appear as either positive or negative, providing a "metaphor for how so many species are on the edge of extinction and how we as humans are entangled in this great loss".[2] It was impossible to look at these daguerreotypes without seeing oneself reflected in the image, inviting observers to reflect on their own positionality and inseparability from the driving processes of extinction. Photographer Denis Defibaugh engaged in a similar practice by taking portraits of specimens of extinct species with a Polaroid Type 55 film. In Defibaugh's understanding, the medium was also the message, as he only had a number of boxes of this (now obsolete) film left and the resulting prints are intended to oxidise over time, thereby creating a daguerreotype effect. The "duality of presentation and decay", Defibaugh noted, "is at the crux of these photographs".[3]

Photographic images occupy the liminal space between here-and-there, now-and-then. As articulated by Elisabeth Roberts, "the pictorial image draws a bounding line, capable of fixing and stabilizing, but within it is the potential for movement" (379). Ella Mudie notes that the common trope of extinction photography is in its performance of "*memento mori*" that functions as means of memorialising past loss (22). Mudie calls for interventions in photography which move beyond mourning and melancholia and a "deathly fixation with the worlds we have already lost" to "focus toward the world we want, or need, to become" (27). Photographs make connections over spatiotemporal scales and across the ontological border of life and death, and thus could be considered "distance reduction technologies" (Despret "Inquiries" 237). They allow for extinction to be affectively and materially engaged, sensed, and conjured in the present (Zylinska 67), despite ghosts of the past haunting futures-to-come.

Another instance of artistic engagement with the photographs of extinct species is *Animal Existence*,[4] by visual artist Brandon Jacob Hudson (2017). Hudson reworks and distorts old photographs of extinct animals such as the thylacine shown in Figure 8.2. This work comes in the context of photographs claiming to show surviving thylacines in Tasmania circulating widely online, whilst cloning and gene editing projects propose to "resurrect" the "extinct" species (Spurling 67). Through a technique of reformation, archival photographs of these species are printed on acrylics which prevent the ink setting as intended, allowing the artist to manipulate the image and experiment with forms of representation. The images are haunting: past photographic representations that materialise in uncanny forms through artistic intervention, posing questions regarding human relations with other species (Searle "Absence" 168–169). Species depicted in *Animal Existence* are partially unrecognisable; they are encountered in distinct historical and cultural contexts and imagined as alternative configurations.

Figure 8.2 Thylacine, from Brandon Jacob Hudson's series *Animal Existence*. Image reproduced with permission.

Hudson's artwork thus metaphorically illustrates the ways the visual images of extinct species continue to haunt contemporary worlds and reminds us how these photographic recollections are transient entities.

We see photographs as a form of "conjuration": this is the term that Derrida uses throughout *Spectres of Marx* as part of his injunction that we "learn to live with ghosts" (xvii–xviii). Learning to live, for Derrida, cannot come from life alone; rather, it is an ethical endeavour at the border of life/death and presence/absence. In our inflection, we take conjuration to signify an incantatory appeal—a calling forth into being of ghost species, a call that is intended to have action on the world and transform the future. The link between conjuration and the idea that time is, to use Shakespeare's words, "out of joint" (ibid. 1), is an important one in the context of a human-driven extinction crisis: conjuration "makes visible the ways in which different temporalities—whether past, present, or future—are imbricated, reminding us that the past continues to bear down on us, obligating us toward the future" (Pollard 147). Hudson's *Animal Existence* taps into this sense in its reworking an archival photograph of one the last recorded thylacines (Figure 8.2). In this image, Hudson flirts with erasure and debasement by occluding and obscuring the features of the individual animal, while visually highlighting the stripes—the visual icon for this lost species.

Another such conjuration occurs in Alexandra Murphy's exhibition *Animal Afterlives* at the University of Cambridge Museum of Zoology, Cambridge (2021). Here, Murphy explores the photograph's relationship with preservation, representation, life and death, and past and present. The exhibition consisted of photographs of different taxidermy specimen collections in the USA and the UK (Ashby). Murphy recounted to us in an interview that this project was inspired by Freud's concept of the uncanny, which she found prevalent in both taxidermy and photography alike. Both photography and taxidermy utilise organic materials to create an affective atmosphere of life-likeness, yet such materials and media remain subject to ongoing decay. In curating this exhibition in a museum space, juxtaposed with taxidermic specimens with dead animals from around the world, Murphy invites visitors to reflect on the power and craft of imaging practices in how they shape modes of reckoning with species loss in the Anthropocene.

The idea that endangered or disappeared species are *presently absent*, not *lost forever*, is a strong theme running through conservationist art and artistic commemorations of extinction, a sense that their absence is apparitional, and therefore brimming with presence. LaCapra (699) raises this distinction between loss and absence in noting that loss is historical: it has a tense, its events can be narrated, and it refers to specific phenomena. In contrast, absence is narrated in a non-conventional way because it is transhistorical and does not refer to a particular event or phenomena. Drawing on LaCapra, we imagine that the loss of an individual species cannot be fully communicated in isolation from the stories of other species. Instead of a bounded historical narrative of loss, these feelings provoke a wider anxiety about absence that is not grounded in the story of a single species. These feelings of "absence as absence" may never be overcome but, like the haunting events of the past, "must be lived with in various ways" (LaCapra 707). For LaCapra, absence is inherently ambivalent because of "its relation to presence, which is never full or lost in its plenitude but in a complex, mutually marking interplay with absence" (ibid.). This interplay between absence and presence is particularly manifested in photographs which use nostalgic or aged filters to provoke an emotional response to extinction.

A final example of photography that evokes a sense of loss can be found in *Last Seen*, a collection of photographs by Lorraine Turi of the last place extinct species was seen in the USA. Lists of extinct species were "translated into visual context in order to create a photographic reference". The artist suggests "the photographs themselves are data made visible a visceral connection between the lost species and humanity" (Turi). The project focused on animals such as the Labrador duck (last seen in Long Island Sound in 1875), the dusky seaside sparrow (last seen 1987 in Florida), and Bachman's warbler (last seen 2001 in the swamp forests of South Carolina). The ordinariness of the images (they contain trains, ploughed fields, blurry skyscrapers) and the absence of the animals—indeed of any animals—provoke reflection of emptiness, absence, and silence, found in the mundane aspects of quotidian life. Such framing encourages us to experience extinction for what it is: not a simple case of before/after, present/absent, and alive/dead, but rather, an everyday experience felt across borders of species and scales of space and time.

Conclusion

In this chapter, we have argued that some non-human animals—those extinct or believed to be extinct—haunt the boundaries imagined to exist between absence and presence through the unsettling nature of photographic representation. Photography as a form lends itself strongly to such liminality because of the potential for creative responses to the documentary image. Such a work is both evidence of presence (at least when the photograph was taken) and an exploration of absence and transformation. Photography as a medium is unique in its capacity to document absence and presence, to record and memorialise the lives of extinct species, and to evoke emotional responses to the loss of living diversity. Photographs conjure ghosts of extinct species and make them present. They occupy a fundamental space in memorial culture, animated and transformed across geographies and histories. They speak to the moral debate about extinction and human thriving that is agreed to be one of the core dilemmas of the Anthropocene.

Notes

1 Smithsonian Museum of Natural History, "Martha, the last passenger pigeon", https://naturalhistory.si.edu/research/vertebrate-zoology/birds/collections-overview/martha-last-passenger-pigeon
2 https://home.dartmouth.edu/news/2017/05/next-kin-seeing-extinction-through-artists-lens
3 https://www.viewarts.org/exhibitions/past-exhibitions/denis-defibaugh-afterlives-of-natural-history
4 https://www.brandonjacobhudson.com/animal-existence

Works Cited

Adams, William M. *Against Extinction: The Story of Conservation*. Earthscan Books, Oxford, 2004.

Adams, William M. "Geographies of conservation II: Technology, surveillance and conservation by algorithm." *Progress in Human Geography*, vol. 43, 2017, pp. 337–350.

Alvey, Mark. "The cinema as taxidermy: Carl Akeley and the preservative obsession." *Framework: The Journal of Cinema and Media*, vol. 48, no. 1, 2007, pp. 23–45.

Ashby, Jack. "Animal Afterlives: Photography, Dioramas, and Forgetting that Taxidermy is Dead." The Natural Sciences Collections Association (NatSCA), 12 December 2021. Accessed 22 August 2022. https://natsca.blog/2021/12/02/animal-afterlives-photography-dioramas-and-forgetting-that-taxidermy-is-dead/

Attenborough, David. *Life on Air: Memoirs of a Broadcaster*. BBC Books, London, 2002.

Bales, Stephen Lyn. "A Close Encounter with the Rarest Bird." Accessed 31 August 2022. https://www.smithsonianmag.com/science-nature/a-close-encounter-with-the-rarest-bird-54437868/

Barrow, Mark. *Nature's Ghosts: Confronting Extinction from the Age of Jefferson to the Age of Ecology*. University of Chicago Press, Chicago, IL, 2009.

Barthes, Roland. *Camera Lucida: Reflections on Photography*. Richard Howard trans. Cape, 1981.

Bastian, Michelle. "Whale falls, suspended ground, and extinctions never known." *Environmental Humanities*, vol. 12, no. 2, 2020, pp. 454–474.

Behr, Bernd. "Akeley inside the elephant: Trajectory of a taxidermic image." *Philosophy of Photography*, vol. 7, no. 2, 2016, pp. 43–61.

Berger, John. *About Looking*. Pantheon Books, New York, 1980.

Bezan, Sarah. "Dodo birds and the Anthropogenic wonderlands of Harri Kallio." *Parallax*, vol. 25, no. 4, 2019, pp. 427–445.

Brown, Sara. "Heath hen tops de-extinction list." *Vineyard Gazette*, July 28, 2016.

Cokinos, Christopher. *Hope is the Thing with Feathers: A Personal Chronicle of Vanished Birds*. Warner Books, New York, 2001.

DeLuca, Kevin Michael and Demo, Anne Teresa, "Imaging nature: Watkins, Yosemite, and the birth of environmentalism." *Critical Studies in Media Communication*, vol. 17, no. 3, 2020, pp. 241–260.

Denis Defibaugh: Afterlives of Natural History. View Center for Arts and Culture, 2015. https://www.viewarts.org/exhibitions/past-exhibitions/denis-defibaugh-afterlives-of-natural-history

Derrida, Jacques. *Specters of Marx: The State of the Debt, the work of Mourning, and the New International*. Peggy Kamuf trans. Routledge, London, 2006.

Despret, Vinciane. "Afterword: It is an entire world that has disappeared." *Extinction Studies: Stories of Time, Death, and Generations*, edited by Deborah Bird Rose, Thom van Dooren, and Matthew Chrulew. Columbia University Press, New York, 2017, pp. 217–222.

Despret, Vinciane. "Inquiries raised by the dead." *HAU: Journal of Ethnographic Theory*, vol. 9, no. 2, 2019, pp. 236–248.

De Vos, Ricardo. "Extinction stories: Performing absence(s)." *Knowing Animals*, edited by Laurence Simmons and Philip Armstrong. Brill, 2007, pp. 183–195.

Dunaway, Finis. "Hunting with the camera: Nature photography, manliness, and modern memory." *Journal of American Studies*, vol. 34, 2000, pp. 207–230.

Fitzpatrick, John W. "Ivory-billed woodpecker (*Campephilus principalis*) persists in continental North America". *Science*, vol. 308, no. 5727, 2005, pp. 1460–1462.

Freeman, Carol. *Paper Tiger: How Pictures Shaped the Thylacine*. Forty South Publishing, Cambridge, 2014.

Fuller, Errol. *Lost Animals: Extinction and the Photographic Record.* Bloomsbury, London, 2013.

Gandy, Matthew. "An Arkansas parable for the Anthropocene." *Annals of the American Association of Geographers,* vol. 112, no. 2, 2022, pp. 368–386.

Garlick, Ben. "Cultural geographies of extinction: Animal culture amidst Scottish Ospreys". *Transactions of the Institute of British Geographers*, vol. 44, no. 2, 2019, pp. 226–241.

Garlick, Ben and Symons, Kat. "Geographies of extinction." *Environmental Humanities*, vol. 12, no. 1, 2020, pp. 296–320.

Ginn, Franklin, Beisel, Uli, and Barua, Maan. "Flourishing with awkward creatures: Togetherness, vulnerability, killing." *Environmental Humanities*, vol. 4, no. 1, 2014, pp. 113–123.

Gordon, Avery F. *Ghostly Matters: Haunting and the Sociological Imagination*. University of Minnesota Press, Minneapolis, 1997.

Guasco, Anna. " 'As dead as a dodo': Extinction narratives and multispecies justice in the museum." *Environment and Planning E: Nature and Space*, vol. 4, no. 3, 2021, pp. 1055–1076.

Haraway, Donna. "Universal donors in a vampire culture: It's all in the family: Biological kinship categories in the twentieth century United States." *Uncommon Ground: Toward Reinventing Nature*, edited by William Cronon. New York, NY: W.W. Norton, 1995, pp. 321–366.

Heise, Ursula. *Imagining Extinction: The Cultural Meanings of Endangered Species.* University of Chicago Press, Chicago, IL, 2016.

Ibbotson, Rosie. "Making sense? Visual cultures of de-extinction and the anthropocentric archive." *Animal Studies Journal*, vol. 6, no. 1, 2017, pp. 80–105.

Johnson, Carolyn Y. "Long-extinct heath hen comes to life in archival film." *Boston Globe*, March 7, 2014.

Jørgensen, Dolly. "Endling, the power of the last in an extinction-prone world." *Environmental Philosophy*, vol. 14, no. 1, 2017, pp. 119–138.

Jørgensen, Dolly. *Recovering Lost Species in the Modern Age: Histories of Longing and Belonging.* MIT Press, Cambridge, MA, 2019.

Kember, Sarah. "After the Anthropocene? The photographic for earthly survival." *Digital Creativity*, vol. 28, no. 4, 2017, pp. 348–353.

LaCapra, Dominick. "Trauma, absence, loss." *Critical Inquiry*, vol. 25, no. 4, 1999, pp. 696–727.

Lammertink, Martjan, et al. "Film documentation of the probably extinct imperial woodpecker (*Campephilus Imperialis*)." *The Auk*, vol. 128, no. 4, 2011, pp. 671–677.

Lippit, Akira Mazuta. *Electric Animal: Toward a Rhetoric of Wildlife*. University of Minnesota Press, Minneapolis, 2000.

Ludtke, A. "Yellow-bellied Weasel Spotted in Cambodia." *Khmer Times*, August 28, 2014.

Lyons, Claire, et al. *Antiquity & Photography: Early Views of Ancient Mediterranean Sites.* Getty Museum, 2005.

McCormack, Derek. "Remotely sensing affective afterlives: The spectral geographies of material remains." *Annals of the Association of American Geographers*, vol. 100, no. 3, 2010, pp. 640–654.

McCorristine, Shane and Adams, William A. "Ghost species: Spectral geographies of biodiversity conservation." *Cultural Geographies*, vol. 27. no. 1, 2020, pp. 101–115.

McIntyre, Sally. "A One Minute Radio Silence for Sceloglaux Albifacies." Accessed 29 July 2022. https://vimeo.com/117256152

Meek, Paul D., Ballard, Guy-Anthony, Fleming, Peter J. S., Schaefer, Michael, Eilliams, Warwick and Falzon, Guy. "Camera traps can be heard and seen by animals." *PLoS ONE*, vol. 9 no. 10, 2014, e110832.

Mitman, George. *Reel Nature*. University of Washington Press, Seattle, WA, 1999.

Mudie, Ella. "Beyond mourning: On photography and extinction." *Afterimage*, vol. 44, no. 3, 2016, pp. 22–27.

"Next of Kin: Seeing Extinction through the Artist's Lens." *Dartmouth News*, 2017. https://home.dartmouth.edu/news/2017/05/next-kin-seeing-extinction-through-artists-lens

Pyne, Lydia. *Endlings: Fables for the Anthropocene*. University of Minnesota Press, Minneapolis, 2023.

Reinert, Hugo. "Face of a dead bird—Notes on grief, spectrality and wildlife photography." *Rhizomes: Cultural Studies in Emerging Knowledge*, vol. 23, 2012. http://www.rhizomes.net/issue23/imre

Roberts, Elizabeth. "Geography and the visual image: A hauntological approach." *Progress in Human Geography*, vol. 37, no. 3, 2012, pp. 386–402.

Rowcliffe, Marcus and Carbone, Chris. "Surveys using camera traps: Are we looking to a brighter future?" *Animal Conservation*, vol. 11, no. 3, 2008, pp. 185–186.

Ryan, James R. "Hunting with a camera: Photography, wildlife and colonialism in Africa." *Animal Spaces, Beastly Places: New Geographies of Human-Animal Relations*, edited by Chris Philo and Chris Wilber. Routledge, London, 2000, pp. 205–222.

Ryan, John C. "Why do extinctions matter? Mourning the loss of indigenous flora in the southwest of Western Australia." *Philament*, vol. 15, 2009, pp. 51–80.

Searle, Adam. "Absence." *Environmental Humanities*, vol. 12, no. 1, 2020, pp. 167–172.

Searle, Adam. "Hunting ghosts: On spectacles of spectrality and the trophy animal." *Cultural Geographies*, vol. 28, no. 3, 2021, pp. 513–530.

Searle, Adam. "Spectral ecologies: De/extinction in the Pyrennes." *Transactions of the Institute of British Geographers*, vol. 47, no. 1, 2022, pp. 167–183.

Sklair, Leslie. "Media coverage of Anthropocene-related creative arts." *The Anthropocene in Global Media: Neutralizing the Risk*, edited by Leslie Sklair. Routledge, London, 2021, pp. 232–251.

Sontag, Susan. *On Photography*. Rosetta, New York, 2005.

Spurling, Amy. "The hunt for the Thylacine." *Engineering & Technology*, vol. 13, no. 5, 2018, pp. 66–69.

Tanner, James T. *The Ivory-Billed Woodpecker*. National Audubon Society, 1942.

Tsing, Anna, et al., eds. *Arts of Living on a Damaged Planet: Ghosts and Monsters of the Anthropocene*. University of Minnesota Press, Minneapolis, 2017.

Turi, Lorraine. "Last Seen". Accessed 30 July 2022. https://www.lorraineturi.com/about-1

Van Dooren, Thom. *Flight Ways: Life and Loss at the Edge of Extinction*. Columbia University Press, New York, 2014.

Wegge, Per, Pokheral, Chiranjibi P. D. and Jnawali, Shant Raj, "Effects of trapping effort and trap shyness on estimates of tiger abundance from camera trap studies." *Animal Conservation*, vol. 7, no. 3, 2004, pp. 251–256.

Wylie, John. "The spectral geographies of W. G. Sebald." *Cultural Geographies*, vol. 14, no. 2, 2007, pp. 171–188.
Yusoff, Kathryn. "Aesthetics of loss: Biodiversity, banal violence and biotic subjects." *Transactions of the Institute of British Geographers*, vol. 37, 2012, pp. 578–592.
Zimmer, Carl. "Century after extinction, Passenger pigeons remain iconic—and scientists hope to bring them back." *National Geographic*, August 31, 2014.
Zylinksa, Joanna. "Photography after extinction." *After Extinction*, edited by Richard Grusin. University of Minnesota Press, Minneapolis, 2018, pp. 51–57.

9 An Elegy for an Ecotype

Eva Saulitis's *Into Great Silence* and the Extinction of the Chugach Transient Killer Whales

Jennifer Schell

On March 24, 1989, the tanker vessel *Exxon Valdez* grounded on Bligh Reef, spilling 10.8 million gallons of crude oil into Alaska's Prince William Sound. In *Into Great Silence: A Memoir of Discovery and Loss Among Vanishing Orcas*, Eva Saulitis, a marine biologist and nonfiction writer, describes both the immediate and lasting ramifications of this disaster. At the beginning of the book, she records the sudden shock of the catastrophe, the relentless spread of the oil, and the first futile efforts at remediation. She also recounts the numerous fatalities caused by the spill, including the deaths of nine of the twenty-two killer whales belonging to the Chugach transient (AT1) ecotype.[1] Throughout the remaining chapters, Saulitis enumerates the enduring effects of the oil on the Sound and documents the long, slow decline of the surviving cetaceans, "staying with the trouble"—Donna Haraway might put it—as to bear witness to their inevitable extinction (1).

While its title declares it a memoir, *Into Great Silence* also represents an ecological prose elegy, which laments both the loss of the unbefouled Sound and its nonhuman inhabitants. Much like others of its kind—Edward Abbey's *Desert Solitaire: A Season in the Wilderness* and Gretel Ehrlich's *The Future of Ice: A Journey Into Cold*—the book chronicles the anthropogenic destruction of entire ecosystems, memorialising their former singularity and beauty for posterity. Similar to Aldo Leopold's address "On a Monument to the Passenger Pigeon" and the poems collected in Karla Linn Merrifield's *The Dire Elegies* and Joanna Lilley's *Endlings*, it also commemorates the lives of those nonhuman animals who have been doomed to extinction by anthropogenic environmental problems, such as pollution, exploitation, habitat loss, and climate change.

Elegies for lost ecosystems have a long history in the Western cultural imagination, largely because as ecocritic Timothy Morton explains, "elegy's formal topics and tropes are environmental" (252). In the European context, they date to the classical era, but in the North American context, they began to emerge in the early nineteenth century. As settlers infiltrated and developed the interior of the continent—razing forests, draining wetlands, and exhausting soils as they did so—some authors sought to mourn and memorialise lost environments by composing elegies for them. Examples

DOI: 10.4324/9781003315957-10

include James Fenimore Cooper's *The Pioneers* (1823), Thomas Cole's "The Lament of the Forest" (1841) and Lydia Huntley Sigourney's *Scenes in My Native Land* (1844), all of which bemoan the rampant deforestation occurring in the rural regions of what is currently New York. Over time, as anthropogenic environmental problems in North America increased, ecological elegies proliferated, recording the destruction of all manner of land and seascapes across the continent. After the onset of the modern environmental movement in the 1960s, their tone increased in intensity and their popularity surged with writers and readers alike.

In Euro-American and Euro-Canadian literature, elegies for specific species have a shorter provenance than elegies for particular places, largely because extinction was not recognised as a scientific fact until the 1820s.[2] Early texts about the disappearance of species emphasise not their decline but their removal to new places. Examples include William Cullen Bryant's "The Prairies" (1833), George Ruxton's *Adventures in Mexico and the Rocky Mountains* (1847), and Herman Melville's *Moby-Dick* (1851), all of which describe nonhuman animals as migrating to new locales in response to anthropogenic hunting pressures.[3] Not until the twentieth century did North American authors become concerned enough about extinction to compose elegies for endangered, extirpated, and exterminated species. Much like those that addressed ecosystems, these elegies also grew more numerous and popular over time, their subject matter resonating with environmentalist readers concerned about diminishing wildlife populations.

Though the archive of North American ecological elegies is vast, *Into Great Silence* is particularly remarkable for its author's willingness to dwell in what Deborah Bird Rose calls the "death zone", or "the place where the living and the dying encounter each other in the presence of that which cannot be averted. Death is imminent but has not yet arrived" (3–4). According to Rose, the death zone calls those who would venture into it "into an ethics of proximity and responsibility" whose "expression will be visible in…how we call out, how we refuse to abandon others, how we refuse hard-heartedness, and thus how we embrace the precious beauty that permeates the house of life" (4). For Rose, this ethics requires "a kind of crazy love that is directed toward earth life", one that "keeps calling others back from the edge of disaster, and staying with those who grieve in the wake of death" (5–10).

As I argue, Saulitis takes readers straight into the death zone to illustrate the dreadful toll of the oil spill on southern Alaska's littoral ecosystems. Then, she proceeds to demonstrate her crazy love for and ethical commitment to the region's killer whales. Instead of leaving after the conclusion of the clean-up efforts—like most media professionals, industry insiders, and government officials—she returns to the scene of the devastation each year so as to document the ongoing effects of the disaster on the Chugach transients. As she does so, she tries her best to understand their complex lifeways, recognise their emotional capacities, comprehend their intellectual abilities, and grieve their unpreventable extinction. Through her efforts to stay with the whales and make their lives meaningful, she shows readers how to embrace crazy love and dwell in the death zone along with suffering cetaceans, devastating loss, and overwhelming grief. As Saulitis, herself, puts it, she illustrates how to "repay with our lives some measure of the love we feel" (101).

Another noteworthy aspect of *Into Great Silence* is the fact that it focuses not on a species or subspecies, but on an ecotype, a genetically distinct, geographically isolated population that possesses the potential to evolve into a new species or subspecies under favourable conditions. Insofar as it elegises the latter and not the former, I argue that *Into Great Silence* subverts the pre-eminence of species and the hierarchy of grievability inherent in contemporary anti-extinction writing. To do so, it employs a complex mixture of discursive strategies, some of which involve scientific discoveries about the Chugach transients, while others address multispecies entanglements and personal connections. Thus, Saulitis explains that these whales possess their own culture: kinship relationships, hunting methods, dietary preferences, and communication systems. Adopting a decolonial approach, she highlights their significance to Alaska Native peoples, likening the extinction of acoustic dialects and Indigenous languages.[4] And she describes her affinity for the region and its whales, especially one individual adult male named Eyak. Taken altogether, this combination of scientific evidence, anthropological information, and personal sentiment provides compelling support for the idea that ecotypes are just as grievable—and worth elegising—as species. It also demonstrates that ecological writing represents an important site of contemporary memorial culture, especially with respect to the anthropogenic extinction of nonhuman animal populations.

Elegies: Ecological and Otherwise

As a form, elegies date to classical antiquity and the works of poets such as Theocritus and Virgil. Composed in strict elegiac meter—alternating lines of hexameter and pentameter—they address a wide array of subjects, including both love and loss (Kennedy 3). Some of these texts adopt the form of the pastoral elegy, an important subvariant of the genre much imitated by later poets, such as John Milton and Percy Bysshe Shelley. More famous and popular than others of their kind, these elegies discuss death and loss in the context of idyllic rural life. Often featuring shepherds, they possess a number of other conventions insofar as many of them allude to classical figures, depict nature as mourning, question the "justice of fate", and provide resolution in consolation (Abrams 77–78).

As literary critic David Kennedy explains, elegy became "strongly identified with a poetry of mourning" by the eighteenth century (4). Although constrained by their subject matter, the poets of this period abandoned elegiac meter and explored new formal options. As they did so, they composed personal elegies that served as memorials for deceased individuals and impersonal elegies that served as meditations on human mortality. An example of the former, Walt Whitman's "When Lilacs Last in the Dooryard Bloom'd" (1865)—written for Abraham Lincoln—represents a translation of the traditional pastoral elegy into extended free verse. An example of the latter, Thomas Gray's "Elegy in a Country Churchyard" (1751) contemplates the speaker's apprehensions about the inevitability of death in a series of elegiac stanzas.[5]

According to John Vickery, elegy further "diversified in form, theme, and attitude" across the twentieth century as poets attempted to come to terms with the various types of "personal, intellectual, and cultural loss suffered by mankind" (1–2). Experimenting with form and content, they composed elegy after elegy—in both poetry and prose—to commemorate the losses incurred by two world wars, as well as the Great Depression, Holocaust, Atomic Era, and AIDS epidemic (Vickery 2–5). These texts certainly push the limits of the elegiac mode. For example, Sylvia Plath's "Daddy" (1965) expresses anger in addition to sorrow. Meanwhile, Thom Gunn's "The Man with the Night Sweats" (2009) and Reginald Dwayne Betts' "Elegy Ending with a Cell Door Closing" (2010) elegise individual people as a means of addressing public health problems and social justice issues.

In addition to their diversity, modern elegies are also marked by their refusal of consolation. Instead of moving towards resolution and detachment, they tend to linger in what Sigmund Freud in his influential essay "Mourning and Melancholia" describes as melancholia or "pathological mourning" (250). Over the last several decades, scholars interested in modern elegiac writing have attempted to complicate and counter the claims about the mourning process that Freud advances in this piece. Combining terms, Jahan Ramazani argues that many modern elegists engage in "melancholic mourning" or "mourning that is unresolved, violent, and ambivalent" (4). Not insignificantly, David Eng and David Kazanjian regard this kind of mourning as productive—not pathological—insofar as it perpetuates an "ongoing and open relationship with the past" that enables "new perspectives on and new understandings of lost objects" (4). For R. Clifton Spargo, it represents a "dissenting act, a sign of an irremissible ethical meaning" that helps to explain why so many elegies describe and discuss problems of social justice (6).

Though insightful and compelling, this scholarship only addresses anthropocentric elegies or those that address human losses. This bias is a peculiar one. After all, over the course of the last two centuries, North American authors have produced myriad elegies for lost ecosystems and species. To rectify this oversight, some ecocritics—Timothy Morton and Bonnie Costello, among them—have set out to describe and discuss the characteristics of these texts. In "The Dark Ecology of Elegy", Morton contends that anthropocentric and ecocentric elegies differ in that the former mourn for "that which has already passed" and the latter lament "that which *will have passed* given a continuation of the current state of affairs" (254). In "Fresh Woods: Ecology and Elegy Among the Ruins", Costello expands upon this idea, characterising ecological elegies as anticipatory or proleptic texts that engage in resistant mourning and environmental activism. As she puts it, "ecoelegy mourns a particular loss...as if it could stop the loss and offer belated protection" (330).

Perhaps not surprisingly, given the severity of the current crises, much of the scholarship on ecological elegies has focused on assessing their capacity to inspire action and activism. Though some scholars see potential in these texts, many more are critical of them. In *Imagining Extinction: The Cultural Meanings of Endangered Species*, Ursula Heise argues that elegiac stories are "both galvanizing and problematic: the nostalgia they generate has often successfully mobilized support for conservation and for critiques of modernization, even as it has made the understanding of ecosystem function more difficult and forward-looking perspectives more inaccessible" (14). Building on Heise's claims, Elspeth Tulloch argues that ecological elegies "may also inhibit action. They may paralyzingly wrap the end into now, heralding the absolute despair of a vague future time" (37). Will Elliott advances a similar claim, positing that "although elegiac representations of Alaska may be effective in the short term at mobilizing audiences in response to specific threats like arctic warming, in the long term they also imply the futility of those responses" (54–55).

While these concerns possess merit, evidence from *Into Great Silence* suggests that ecological elegies may not be as paralysing or ineffectual as these scholars indicate. In her prologue, Saulitis reminisces about watching *The Last of the Curlews* (1972)—an ABC Afterschool Special adapted from the book by Fred Bodsworth—as a nine-year-old girl. She describes the experience as a powerful and affecting one, an "origin moment" which helped her to understand "extinction not with my mind but my heart" (xiii–xiv). Instead of inspiring apathy or depression, this ecological elegy prompted Saulitis to care about and grieve for the loss of endangered birds and other nonhuman organisms. Significantly, these emotions did not dissipate over time. As she, herself, explains, "I can still summon my ache for that last curlew; almost forty years later, it throbs in my chest" (xiii). Needless to say, Bodsworth's elegy—and its animated adaptation—did not deter Saulitis from earning a degree in marine biology and studying the Chugach transients. Neither did it prevent her from staying with and writing about them so as to mourn and memorialise their lives. In the end, then, Saulitis demonstrates that elegies for endangered species can be more inspirational than incapacitating for scientists and creative writers, alike.

Here, I want to add that judging ecological elegies according to their ability to inspire activism can be problematic, because this is not necessarily their only or even their primary function. Generically speaking, all elegies—ecological variants included—help writers and readers to confront and contend with the grief they experience when they suffer a loss. They make space for people to sit with their emotions. They may not offer consolation or resolution, but they commemorate the dead, making continued relationships with and new understandings of them possible. Part of what makes the scholarly insistence on the association of ecological elegy and human agency troubling is that it ignores and dismisses the value of these other functions. Insofar as it promotes the idea that human beings can save the planet if they take action to do so, it also requires a willingness to remain dedicated to ideas about human exceptionalism. And last, but not least, it reduces a complex, beautiful art form to a simple, motivational tool for environmental activists.

Instead of evaluating ecological elegies solely according to their ability to inspire action, I suggest that ecocritics consider their capacity to help humans cope with the grief caused by living in an era of anthropogenic climate change and mass extinction. After all, as Ashlee Cunsolo and Karen Landman explain in their introduction to the edited collection *Mourning Nature: Hope at the Heart of Ecological Loss and Grief*:

> We are entering a time when ecologically based mourning seems likely to occupy more and more of our experience. And yet it seems that we are also entering a time of great denial and avoidance of this type of work because of what it means and what it will entail if we were to truly embrace and open ourselves to these changes and this mourning, and to the understanding of our individual and collective responsibility.... How do we even begin to think and feel *that*?
>
> (6)

Though Cunsolo and Landman go on to furnish some possible answers to the question that they pose at the end of this passage, I want to suggest that we begin by embracing elegies, because they have long played an integral role in anthropocentric memorial culture. Ever since the classical era, they have helped humans commemorate their losses, explore their emotions, and grapple with their grief. And, as *Into Great Silence* demonstrates, they can serve a similar function in ecocentric memorial culture.

Cetacean Science: Inspiring Care and Elegising Orcas

As I noted earlier, Saulitis employs a number of distinctly different but deeply imbricated discursive strategies in order to memorialise the killer whales of Prince William Sound. The first of these involves scientific research on cetaceans. Over the course of the book, Saulitis adopts an ethnographic approach towards the Chugach transients, encouraging readers to relate to and care for them by highlighting their familial structures, acoustic dialects, dietary preferences, predatory habits, and reproductive struggles. In these passages, she adeptly navigates the fraught terrain between anthropomorphism—attributing human characteristics to nonhuman animals—and anthropodenial—denying the "shared characteristics between humans and animals" (De Waal 258). Finishing the book with a flourish, she expands her scope beyond the Sound, describing different orca ecotypes from around the world and explaining why each one deserves appreciation and protection from humans.

Importantly, Saulitis only addresses the Chugach transients, not the other two killer whale ecotypes that live in the Gulf of Alaska, the residents and the offshores (5). This strategy serves to distinguish her book from most other ecological elegies for endangered or extinct organisms, which tend to focus on species or subspecies. One reason why the latter have been so popular with environmentalist writers is because they represent basic taxonomic categories that help humans organise and understand the natural world. To put it another way, they constitute quantifiable units that authors and activists can use to measure the extent of anthropogenic destruction enacted upon the natural world. They also represent distinct entities that humans can grieve when they vanish from the planet. Not insignificantly, some of them—the dodo, thylacine, and passenger pigeon—possess the capacity to serve as icons, important symbols of human rapacity towards and carelessness for nonhuman organisms.

Here, it is important to acknowledge that despite its usefulness for environmentalists, species is an imprecise and problematic term. In *Defining Species: A Sourcebook from Antiquity to Today*, philosopher John S. Wilkins explains that species is a culturally determined category that emerged in the classical period and evolved over time. After tracing this lengthy history and enumerating countless definitions, he lists the twenty-six different species concepts currently in use by scientists (193–198). Meanwhile, in *Recovering Lost Species in the Modern Age: Histories of Longing and Belonging*, historian Dolly Jørgensen summarises the issues involved, noting, "The whole concept of species itself is a historical construction, with unclear delineations between what constitutes a species versus a subspecies" (9). Though *Into Great Silence* does not directly address these definitional problems, it revolves around a particular ecotype not a single species; thus, it undermines the pre-eminence of the latter in contemporary anti-extinction writing. Insofar as it demonstrates that ecotypes are just as grievable as species, it also suggests that humans should extend their care and concern to all animals affected by anthropogenic ecological disasters, regardless of their population group or taxonomic category.

Scientifically speaking, Saulitis uses the early portions of *Into Great Silence* to describe the essential characteristics of the Chugach transient ecotype, informing readers that they constantly roam throughout the Sound in search of their preferred prey, marine mammals (5). She adds that they live in a "complex society" with the older females serving as "matriarchs, the locus around which daily life revolves" (49). And she argues that they possess culture, because they transmit certain ecological knowledges, navigational abilities and hunting techniques from one generation to another.

According to Saulitis, "Hunting habits and routes are another aspect of culture, another tradition. Calves learn to navigate the Labyrinth by following adults. They learn seal haul-out rocks, points where Dall's porpoises feed. They learn silence" (83). Lest her readers remain unconvinced, she advances a compelling metaphor in which she compares the orcas to humans—both Indigenous Alaskans and European immigrants—and she concludes that "Food is place; food is culture. For us, for whales. The way an animal hunts, the way a human hunts, is culture" (84).

Possessing special insight into cetacean communication because of her scientific research, Saulitis explains that each orca ecotype possesses its own acoustic dialect, which it uses for the purposes of identification and correspondence. According to her, "These dialects reflect cultural traditions", because "Calves learn calls from adults around them" (5). After spending several summers with the Chugach transients recording their conversations, Saulitis discovers that they share "no calls with other populations, suggesting genetic isolation" (197). What this means is that when the whales go extinct, so will their acoustic dialect. Focusing specifically on matters of culture and geography, Saulitis poetically describes the potential impact of these losses on the Sound:

> The Chugach transients were a manifestation of place, more resident to the Sound than most of the whales we called residents. If language is a reflection of place, as linguists claim, then the language of the Chugach transients, both acoustic and behavioural – no, *everything* about them, including their dying—was a reflection of place.
>
> (174)

In addition to describing various aspects of killer whale culture, Saulitis spends some time debunking prevailing stereotypes that position these animals as mindless predators and vicious carnivores. Towards the end of the first part of the book, she describes an instance in which she witnessed three male Chugach transients swimming alongside and playing with one another. Surprised by their amicable interactions, she admits, "I hadn't imagined transients as relaxed animals. I'd imagined them as dogged, secretive predators, aloof, continually on the hunt. I'd imagined so little of what they were" (49). If this moment prompts Saulitis to reflect upon her personal biases, then it also it inspires her to recognise that the orcas possess emotions and individuality. As she says, "Far more than eating machines, they were, like us, creatures of moods and personalities. Like wolves and lions, like humans, they could be as gentle and playful as they could be fierce" (49). Here, as it does elsewhere in the book, the comparison to humans serves not to anthropomorphise the whales but to acknowledge their cognitive complexity. It also helps readers to relate to them.

Significantly, *Into Great Silence* follows Saulitis through twenty-four field seasons, showing her willingness to stay with the Chugach transients as their population numbers steadily decline. She records fifteen deaths and no births. As she does so, she shifts the focus of her research, turning her attention away from recording acoustic dialects and towards collecting skin samples. After sending these biological specimens to a laboratory for chemical testing, Saulitis learns the degree to which harmful, human-made pollutants accumulate in cetacean bodies:

> In terms of PCBs, PBDEs (flame retardants), and DDTs – all immune suppressors and endocrine disruptors – transient orcas, at the top of the marine mammal food chain, are some of the most contaminated creatures on earth. Mothers pass contaminants to their calves through their milk. The poisons originate from leaking barrels dumped decades ago into the Gulf of Alaska by the military. Or they arrive within air masses from Southeast Asia, warm air hitting the cold air over the Gulf of Alaska, raining the toxins down.
>
> (211)

She later concludes that the remaining Chugach transients are probably "too contaminated…to reproduce successfully" (221). In an attempt to come to terms with these devastating but unsurprising results, Saulitis celebrates the surviving whales in striking prose that endows them with vibrant emotional and metaphorical significance: "Those seven, it's true, have been deprived, in the biological sense, of potency, the capacity to reproduce. But they exist, as fully alive as ever, and they pulse with the potency of last ones" (236).

At the very end of the book, Saulitis returns to the question of ecotypes. What makes this passage so important to mark is the fact that orcas are not necessarily an endangered species. Rather, they exist around the world in geographically scattered populations, some of which are more threatened than others.[6] Referring to a poster entitled, "Killer Whale Ecotypes and Forms", Saulitis makes the case for the existential importance of each individual group:

> It illustrates ten distinct orca types now known to exist in the world, six of them in the Antarctic. Each ecotype occupies the ocean in a unique way, from the 'carousel feeding' herring eaters of the North Atlantic, to seal hunters of the Antarctic, who work as a group to create pressure waves to wash their prey off ice floes. There are penguin eaters, shark eaters, and minke whale eaters. One mammal-eating type, named 'Bigg's killer whale', includes the Chugach transients. Like nested Russian dolls, ever more intricate stories reside within those ecotypes, the stories told by those skin samples. Within an ecotype are separate populations, each with its own narrative. Within Bigg's killer whale are West Coast transients, Gulf of Alaska transients, Eastern Aleutian transients, and Chugach transients. All stocky, tall-finned, black-and-white, nonetheless each population lives an entirely different lifestyle—and has for thousands of years.
>
> (211)

In addition to underscoring the sheer diversity of orca ecotypes, this passage showcases Saulitis' artistic talent, namely her ability to invest minimalist, expository prose with intricate metaphors, and intense emotions. Note the way in which she transforms a set of esoteric scientific details about whale behaviour into a series of nested ethnographic stories about cetacean lifeways. In so doing, Saulitis demonstrates that the category of species is just too broad to do justice to the complexity and precarity of these whales. And she renders ecotypes grievable. As she concludes, "There is no one thing called *Orcinus orca*. What we thought one thing is many. What we thought abundant is not. What we thought invulnerable is fragile, existing in that form nowhere else on earth" (211).

Multispecies Entanglements: Lost Languages and Disappearing Dialects

Although she includes a great deal of scientific information in her memoir, Saulitis also spends a hefty amount of time highlighting the connections between the humans and orcas who inhabit Prince William Sound. In other words, she describes the multispecies entanglements in the ecosystems she studies. As she does so, she adopts what feminist science studies scholar Juno Salazar Parreñas calls a decolonial approach, which takes into account the destructive impact of resource extraction on both the Sound's Indigenous groups and whale populations. First, Saulitis describes the arrival of the American salmon fishing fleet—and the canneries that served it—in the Gulf of Alaska; then, she discusses the impact of these historical and environmental events on the Eyak, carefully likening their situation to that of the Chugach transients after the oil spill.

While Parreñas admits that this kind of decolonial approach might seem offensive insofar as it attempts to equate human and nonhuman exploitation and suffering, she contends that dismantling colonialism "requires rejecting the refusal to acknowledge the possibility of shared experiences with nonhuman others" (24). For her, decolonial approaches to the world enable and empower those artists, activists, and scholars who would embrace them, because they make it possible to recognise "that colonialism has brutal impacts for many of Earth's inhabitants, many of whom are not human" (24). In addition to accomplishing this feat, Saulitis' discussion of the effect of extractivist activities on the orcas and the Eyak is insightful because it reveals some of the factors that make certain populations particularly vulnerable to colonialist disruptions.

Throughout the book, Saulitis weaves references to and stories about Indigenous people into her discussions of the Chugach transients. Instead of speaking for Alaska Natives in these passages, she quotes anthropologist Herbert O. Anungazuk (Iñupiat), who explains that some Iñupiaq elders believe that humans "reincarnate as orcas" (166, 228). Meanwhile, the Alutiiq endow killer whales with nonhuman personhood, recognising their ability to steal "someone's spirit" and cause their death (228).[7] For Saulitis, these ethnographic details are important, because they indicate that "the stories of whale and human are so intertwined that they can't be told one without the other" (228). Here, I want to add that they also serve to portray Prince William Sound as a complex place where humans and whales forge enduring relationships, their cultures mingling such that each one affects the other in myriad, mysterious ways.

Perhaps the most perceptive and poignant discussion of multispecies entanglements occurs in the chapter entitled "Lament for Eyak", which grieves both the loss of an individual whale and an Indigenous language. Not coincidentally, both share the same name, Eyak. After explaining that these two extinction stories are inextricably intertwined, Saulitis provides readers with background information about the impact of Euro-American colonisation on the Indigenous groups living in Southern Alaska. Note that in the following passage, she focuses specifically on linguistic issues:

> Devastated by the influx of Europeans to the region, by the late nineteenth century, thirty or so Eyak survived in Cordova, only half of whom still spoke the native language.…Reporter Elizabeth Kolbert, in an article called 'Last Words: A Language Dies', published in the *New Yorker* in 2005, described the Eyak as 'a mysterious people' who lived 'in close proximity to other, more formidable nations', but who for thousands of years 'somehow managed to maintain not only their own culture but also their own language'.
>
> (228)

Because they managed to preserve their cultural traditions while living alongside other populations of people, the Eyak remind Saulitis of the whales that she studies. After making the connection overt for her readers, she indulges in some speculation about the whales, "Perhaps geographical isolation – say, being cut off from other transients during the last ice age – established their distinctness. But for thousands of years since the ice retreated, their isolation from other transients has been maintained culturally, without geographical barriers" (228–229). For Saulitis, the orcas' stubborn fidelity to their lifeways is simultaneously admirable and heart-breaking. As she puts it, "Something kept the Chugach transients faithful to their isolation, even at the cost of survival. They refused to assimilate. It dooms them and at the same time imparts on them dignity" (229).

Although this comparison runs the risk of reinforcing deleterious stereotypes about "vanishing Indians", Saulitis avoids this pitfall by explaining some of the differences between the Eyak and the transients and their responses to settler colonialism and environmental disaster. Astutely, she observes:

> Orcas are not like us. At all costs, they hold true to traditions, like whom to mate with and what language to speak. The surviving Eyak people gradually assimilated, aligning with the powerful Tlingit. They

> adopted new traditions, dropped old ones. Gradually, their language slipped away. Some might say Eyak died as a culture. But others might say it evolved. The culture evolved into a new version of Eyak.
>
> (229)

Simply put, humans possess far more flexibility and resiliency than killer whales. Capable of and willing to adapt to new circumstances, the Eyak remain extant; incapable of or unwilling to adjust to changing conditions, the orcas remain doomed to extinction.

To conclude the chapter, Saulitis presents readers with one of the most moving, elegiac passages in the entire book. First, she confesses her deep emotional attachment to and crazy love for the whales, acknowledging that "Some day, there will be a last Chugach transient" and admitting that "I dread that day" (236). After quoting several particularly poignant lines from Anna Nelson Harry's poem "Lament for Eyak" (1982), she likens extinct human and whale languages:

> After the last one dies, the last Chugach transient, that dialect, like Eyak, will exist only in recorded form, as an artifact. We'll still be able to listen to it, and like a strange poem, wonder at its meaning. Like a poem, it holds a map and a code, the secrets of a way of living in a place. Like a poem, it will remain untranslatable, until we listen close enough to the place to hear and understand its deep language. Once, in the distant time, humans and animals spoke the same language. *That* language.
>
> (237)[8]

Through the use of simile and repetition, Saulitis crafts a beautiful memorial to the Chugach transients that transforms their acoustic dialect into an art form, an eerie and unintelligible poem. While her early sentences express great sadness and grief, she ends on a somewhat more hopeful note, suggesting that humans might someday be able to understand the language of the whales so long as they acknowledge their kinship with nonhuman organisms.

Personal Involvement: Making Connections and Grieving Losses

Throughout the opening chapters of *Into Great Silence*, Saulitis augments her discussions of cetacean science and multispecies entanglement with her personal reflections on the *Exxon Valdez* oil spill and its immediate aftermath. In these early sections of the book, she takes readers on a harrowing journey straight into the death zone to show them the extent of the destruction caused by the accident. Refusing consolation or resolution, she mourns the loss of Prince William Sound's diverse littoral ecosystems and elegises their former fecundity. Over the course of the remaining chapters, Saulitis stays with the trouble, discussing her efforts to study the Chugach transients and document their decline. In these passages, she describes the deeply personal relationships that develop between human and cetacean as she follows the whales across the Sound, recording their calls and observing their behaviours. Importantly, these portions of the book transform the whales from objects of study into subjects to grieve. Insofar as they evidence her crazy love for the orcas, they also show readers how to live ethically in the death zone, witnessing, grieving, and memorialising the losses that result from anthropogenic ecological disasters.

For the most part, Saulitis writes about the spill in retrospect, recalling and relaying her memories from that time as well as those of her friends and colleagues. Significantly, these descriptions are so filled with dreadful detail that they read like the screenplay for an ecohorror film. According to Saulitis, the oil initially "spread in a slow-moving acre from the point of rupture as though uncertain" (60). Then, the wind whipped the "crude into a toxic lather called mousse, spreading it beyond any hope of containment" (60). She adds that some of the first scientists and photographers to enter the disaster zone, "gagged on hydrocarbon fumes" and witnessed "dead animals… in thick windrows, so coated they'd been unrecognizable" (62–63). Later that summer, Saulitis sees the devastation for herself when she encounters "a dead seal washed up on Whale Camp beach, an oil-stained plastic pompom (a clean-up tool) wrapped in our prop, [and] oil pooled ankle deep beneath a boulder" (91). All of these things—though disturbing in their own right—pale in comparison with Saulitis' gruesome discovery of "garbage bags filled to bursting" with the oily bodies of "dead animals" (95).

Sometimes, Saulitis lends immediacy and emotion to her writing by sharing excerpts of her personal writing—her letters, logbooks, and journals. In the following quotation, taken from a letter that she wrote to her parents on 22 June 1989, her outrage and frustration are palpable: "The insidious, idiotic, unthinking, stinking poison has seeped down below the surface. The animals struggle to survive and there is some spark of hope because isn't the beauty and immensity just too powerful to destroy?" (80). Other passages, meanwhile, express feelings of sadness and grief, for they compare weathered hemlock trees to "ancient men in Tai Chi poses" and giant granite stones to extinct sea cows (80). Taken as a whole, then, this letter and its various figures of speech attest to the range of emotions—many of them unbearably intense—that Saulitis experienced as she lived and worked in the death zone in the summer of 1989.

In later chapters, those that describe the first few years after the spill, Saulitis relies less on memories and more on her journals. Insofar as they continue to emphasise the visible and invisible toll of the spill on the ecosystems of Prince William Sound, they refuse all of the consolation and resolution that time and distance have to offer the bereft. After entering the Gulf of Alaska roughly one year after the spill, Saulitis describes the tenacious ubiquity of the oil in her journal, observing that "Even now, after more than a year, tar balls entwine with eelgrass, kelp and twigs in the wrack-line. In rock crevices, oil is wet" (158). As she explains,

> Every living thing I see is a survivor, having come through the spill and then the stress of the winter after. But there's much I can't see: all the oil that remains, buried under gravel, sunk to the sea floor; all the weakened animals; all the animals that didn't make it.
>
> (128)

Though she appreciates all of the extant wildlife she encounters in the Sound, Saulitis resists the urge to celebrate; instead, she continues to mourn the myriad, unknown organisms affected by the spill, both those who perished during the past winter and those who will perish in future seasons.

Eventually, Saulitis turns her attention away from the spill—and its lasting effects on the Sound and its ecosystems—and towards the Chugach transients. In these portions of the book, she introduces the whales by their names, describing their individual characteristics and discussing their intraspecies relationships. Along the way, she reveals the degree to which she comes to care for an individual adult male named Eyak. Initially, this whale plays only a small role in the narrative, appearing briefly in the prefatory materials and the ninth chapter. Then, Saulitis catches him engaging in some remarkably strange behaviour. After spending time swimming in solitude apart from his pod, Eyak starts slapping his fins and flukes on the water. At the same time, he begins vocalising, emitting a strange series of calls that sound like "yawls", "wails", and "blasts" (104). Confused but fascinated by this barrage of aberrant activity, Saulitis contemplates the encounter from a number of different perspectives—her excitement evident in the flurry of questions she poses about cetacean sentience—and she decides to spend her scientific career studying the Chugach transients and trying to unravel their secrets.

Over time, as their interactions increase in frequency and intensity, Saulitis only grows more attached to Eyak. Aboard her research vessel, she records his calls and scrutinises them, trying to locate patterns, make comparisons, and form interpretations. She also documents his solitary journeys through the Sound and records his social interactions with other orcas. As the years pass, she witnesses both playful and peaceful moments, including one in which all the whales—now habituated to her presence—engage in a behaviour she calls "deep rest": "Line-abreast, their bodies touching, they surfaced in synchrony, inscribing a big slow circle. For two and a half hours, this circle drifted from one side of the passage to the other" (152). While this experience arouses her scientific curiosity, it also serves to strengthen her affection for the transients—Eyak included—because she appreciates their willingness to share such an intimate aspect of their lives with her.

Given the intensity of her connection to the whales, it is perhaps not all that surprising that Saulitis is devastated when Eyak grounds and dies on a beach in Prince William Sound. Related in matter-of-fact prose, her articulation of the event is rife with latent, suppressed emotion:

> At 11:00 a.m. on July 11, 2000, Eyak, his breathing labored, swam in circles off the Orca Cannery dock, north of Cordova. Witnesses took photographs. A half hour later, he swam into Hartney Bay, in Orca Inlet, a few miles south of town. There he beached again. Volunteers arrived an hour later, dousing him with saltwater, laying moist canvas on his body to prevent dehydration, hoping he'd swim free during the next high tide. For three hours, people worked to save him. Someone collected sloughed skin for a sample. But despite their efforts, at 4:15 p.m. on July 11, Eyak died. He was thirty-three years old.
>
> (231)

Through these short, staccato sentences, Saulitis manages to relay the bare facts surrounding Eyak's death and convey the care that the volunteers displayed as they worked to alleviate his pain and save his life. Also evident in these lines—and their forced emotional detachment—is the extent of her love and affection for this whale.

Rather than ending *Into Great Silence* with descriptions of Eyak and his death, Saulitis proceeds to describe how the residents of Cordova decided to memorialise him, by preserving his bones and rearticulating his skeleton for display in the Ilanka Cultural Center and Museum. Like other elegists before her, she also attempts to offer some words of solace, consolation, and resolution. Assuming an attitude of peaceful acceptance, she declares,

> I loved the place before I loved the whales. I will still love the place when they're gone. But it will be a different kind of love, and it will be a different place, and I will be a different person.
>
> (240)

Though she desperately wants to believe these sentiments, she soon realises that they are too simple to account for the complexity of her attachment to and the immensity of her grief for the Chugach transients, and she sets out to craft a different ending for her memoir.

In the very last chapter of the book, Saulitis travels to Cordova to pay her respects to Eyak. Though eleven years have passed since his death, she is overwhelmed with emotion as she sits in an adjacent stairwell and gazes at his bones. Perhaps to ease the pain she feels, Saulitis imagines herself speaking to Eyak, telling him, "I'm sorry I wasn't there when you died. I'm sorry it's taken me so long to come here. I miss you in the Sound. Paddy finally has an adult fin. So does Mike" (243). She envisions him responding to her apologies and updates not with comfort and consolation, but with mystical cetacean wisdom. Equating the experiences of human and nonhuman organisms, he says, "Remember, once my ancestors walked on land... Once, Eva, your ancestors swam" (243). In this moment, time collapses and Saulitis visualises herself "crawling inside" Eyak's ribcage,

> curling up in the place where there was once a heart. Suddenly seasick, dizzy, as though Eyak had swept me into his perpetual dive, I sensed how weighted, how wrong, his body must have felt on land. I had to straighten, take deep breaths, look away.
>
> (244)

To make sense of this deeply emotional personal experience, Saulitis considers the trajectory of her life and her time with the whales. First, she recognises the irreparability of their loss and the inevitability of their death:

> I thought of the hole in the heart of the Chugach transients, the deathblow of the spill. No love or science or law or ritual could repair it. Into the hole Eyak and Eccles and Aligo and Holgate had descended. One day, the others would follow. Sooner or later, so would I. Perhaps that's why I sensed falling. Down into the cradle of ribs, where once Eyak's great heart pounded.
>
> (244)

Then, she conceptualises the grieving process as an ongoing one which does not possess a simple or straightforward resolution. For her, though, some solace can be found by engaging in acts of care for the more-than-human world. As Saulitis puts it, meaning and hope—what she calls the "spirit line"—lie in "acting out the heart's one prayer: that what's broken can be mended. That what's shattered can be made whole. That what's damaged can be repaired" (245). Far more realistic and ethical than the one that she elaborates earlier in the memoir, this approach to ecological grief addresses the complex emotional realities inherent in living in an era of profound anthropogenic environmental crisis.

Throughout *Into Great Silence*, then, Saulitis skilfully combines discussions of scientific information and multispecies entanglements with her personal reflections about the *Exxon Valdez* oil spill to create a powerful elegy for and moving memorial to the Chugach transients. To put it another way, she transforms the members of this ecotype into grievable subjects, commemorating their distinct individuality and mourning their inevitable extinction. In the end, it does not matter whether *Into Great Silence* inspires activists to fight for increased ecological protections for Prince William Sound. After all, the book manages to accomplish something just as important: it takes readers to the death zone to show them how to live, love, and grieve ethically in a world experiencing anthropogenic environmental catastrophe. In so doing, it indicates that ecological elegy has an integral role to play in the memorial culture of the Anthropocene.

Notes

1 Scientists refer to this group as the AT1 transient population, and they use corresponding alphanumeric names (AT1, AT2, AT3) to designate its members. Throughout her memoir, Saulitis employs what she calls "familiar names" that indicate the whales' connections to the places and peoples of Prince William Sound (xi). She tends to refer to these cetaceans as orcas—not killer whales—because the moniker is "derived from the name of a Roman god of the underworld" (xii).

2 In "On the Species of Living and Fossil Elephants", Georges Cuvier proposed the extinction of some prehistoric elephant species. Though it was initially rejected by such prominent intellectuals as Thomas Jefferson and Georges-Louis Leclerc, Comte de Buffon, this idea eventually gained the widespread acceptance of the scientific community.

3 Bryant refers to buffalo, while Ruxton describes beaver and Melville addresses whales. Note that "The Prairies" ties the disappearance of nonhuman animals to the disappearance of Indigenous peoples, reinforcing pernicious stereotypes about "vanishing Indians" that served to justify the seizure and settlement of North America by colonial powers (Brantlinger 3).

4 Alaska Native is an inclusive term used to refer to the Indigenous peoples living in what is currently the state of Alaska.

5 An elegiac stanza contains four lines of iambic pentameter that follow the rhyme scheme abab.

6 Some ecotypes, such as the famous Southern Resident group in the Pacific Northwest, have been designated as endangered species, but most others have not. Because much is still unknown about the precarity of orca populations and their degrees of speciation, the International Union for Conservation of Nature (IUCN) refers to the status of the species as "data deficient".

7 The Iñupiat live on the shores of the Chukchi Sea and Arctic Ocean, while the Alutiiq live on the shores of the Gulf of Alaska.

8 The poem is available in its entirety in *In Honor of Eyak: The Art of Anna Nelson Harry*. Saulitis quotes the following lines: "My aunts are dying off on me and alone I'll be living/Why, I wonder, are these things happening to me?/My uncles also have all died out on me and I can't forget them" (155–157).

Works Cited

Abrams, Meyer Howard. *A Glossary of Literary Terms*. 8th ed. Thomson Wadsworth, Belmont, CA, 2005.

Brantlinger, Patrick. *Dark Vanishings: Discourse on the Extinction of Primitive Races, 1800–1930*. Cornell University Press, Ithaca, NY, 2003.

Costello, Bonnie. "Fresh Woods: Elegy and Ecology among the Ruins." *The Oxford Handbook of the Elegy*, edited by Karen Weisman, Oxford University Press, Oxford, 2010, pp. 324–342.

Cunsolo, Ashlee and Karen Landman. "Introduction: To Mourn Beyond the Human." *Mourning Nature: Hope at the Heart of Ecological Loss and Grief*, edited by Ashlee Cunsolo and Karen Landman, McGill-Queen's University Press, Montreal, 2017, pp. 3–26.

De Waal, Frans. "Anthropomorphism and Anthropodenial: Consistency in Our Thinking about Humans and Other Animals." *Philosophical Topics* vol. 27, no. 1, 1999, pp. 255–280.

Elliott, Will. "Raven's World: Ecoelegy and Beyond in a Changing North." *Critical Norths: Space, Nature, Theory*, edited by Sarah Jaquette Ray and Kevin Maier, University of Alaska Press, Fairbanks, AK, 2017, pp. 53–67.

Eng, David and David Kazanjian. "Introduction: Mourning Remains." *Loss: The Politics of Mourning*, edited by David Eng and David Kazanjian, University of California Press, Berkeley, CA, 2003, pp. 1–25.

Freud, Sigmund. "Mourning and Melancholia." *The Standard Edition of the Complete Psychological Works of Sigmund Freud*, edited by James Strachey, vol. xiv, Hogarth, 1957, pp. 243–258.

Haraway, Donna J. *Staying with the Trouble: Making Kin in the Chthulucene*. Duke University Press, Durham, NC, 2016.

Harry, Anna Nelson. "Lament for Eyak." *In Honor of Eyak: The Art of Anna Nelson Harry*, edited by Michael E. Krauss, Alaska Native Language Center, 1982, pp. 155–157.

Heise, Ursula K. *Imagining Extinction: The Cultural Meanings of Endangered Species*. University of Chicago Press, Chicago, IL, 2016.

Jørgensen, Dolly. *Recovering Lost Species in the Modern Age: Histories of Longing and Belonging*. MIT Press, Cambridge, MA, 2019.

Kennedy, David. *Elegy*. Routledge, London, 2007.

Morton, Timothy. "The Dark Ecology of Elegy." *The Oxford Handbook of the Elegy*, edited by Karen Weisman, Oxford University Press, Oxford, 2010, pp. 251–271.

Parreñas, Juno Salazar. *Decolonizing Extinction: The Work of Care in Orangutan Rehabilitation*. Duke University Press, Durham, NC, 2018.

Ramazani, Jaham. *Poetry of Mourning: The Modern Elegy from Hardy to Heaney*. University of Chicago Press, Chicago, 1994.

Rose, Deborah Bird. "In the Shadow of All This Death." *Animal Death*, edited by Jay Johnston and Fiona Probyn-Rapsey, Sydney University Press, Sydney, 2013, pp. 1–20.

Saulitis, Eva. *Into Great Silence: A Memoir of Discovery and Loss among Vanishing Orcas*. Beacon, Boston, MA, 2013.

Spargo, R. Clifton. *The Ethics of Mourning: Grief and Responsibility in Elegiac Literature*. Johns Hopkins University Press, Baltimore, MD, 2004.

Tulloch, Elspeth. "Whose Arctic? Who Cares? Place, Responsibility, and Elegiac Purpose in the Eskimo Curlew Extinction Narrative." *Critical Norths: Space, Nature, Theory*, edited by Sarah Jaquette Ray and Kevin Maier, University of Alaska Press, Fairbanks, AK, 2017, pp. 15–51.

Vickery, John B. *The Modern Elegiac Temper*. Louisiana State University Press, Baton Rouge, LA, 2006.

Wilkins, John S. *Defining Species: A Sourcebook from Antiquity to Today*. Peter Lang, Pieterlen and Bern, 2009.

10 Lost Species, Lost Worlds

Memorialising Extinction in the Art of Todd McGrain and Chris Jordan[1]

Hannah Stark

Memorial culture is a public culture, shaping and shaped by public and collective feelings. This chapter considers art that memorialises and raises awareness about species loss and diminishment. It examines art as offering an affective engagement with extinction politics and a complex nexus point for processing environmental grief and shame, memory and forgetting, and human complicity in environmental destruction in the Anthropocene. This chapter brings together two very different series of artworks about birds, extinction, and endangerment: Todd McGrain's large bronze sculptures of extinct birds, placed and encountered in the location in the United States in which they were last seen in the wild as part of *The Lost Bird Project*; and Chris Jordan's *Midway Atoll* series of photographs of decaying bird remains and the plastics they have imbibed. Both projects speak to profound loss and create opportunities for mourning extinction in the Anthropocene.

Memorial culture reflects our values back to us. How and what we memorialise is political and speaks to cultures of loss and memory. Erika Doss articulates that "monuments and memorials are memory aids: materialist modes of privileging particular histories and values" (38). The memorialisation of extinction, in particular, tells us something about the place of non-human life and death in our culture. Cultural responses to extinction can reveal the pervasiveness of anthropocentrism in the Anthropocene, as well as the modes through which it is critiqued. "In examining how a species lives in memory after its death", Kelly Enright asserts, "cultural conversations about nature and wildness are revealed, and about the placement of animals within stories of human history" (155). At the same time, what we memorialise and the stories we tell ourselves about extinction contribute to our understanding of the impact of the human on more-than-human lifeworlds.

Extinction is, of course, a natural process and is part of the ebb and flow of evolution. As Joshua Schuster writes, the "extinction of species is one way we have come to understand both vast stretches of time past and the precariousness of life today" (88). As the condition we now find ourselves in, precariousness marks the extinction crisis (or sixth extinction event) that is now underway, with humans widely acknowledged as the accelerant. The 2019

DOI: 10.4324/9781003315957-11

Intergovernmental Science-Policy Platform on Biodiversity and Ecosystem Services (IPBES) *Global Assessment Report on Biodiversity and Ecosystem Services* notes that extinctions are "tens to hundreds of times higher than it has averaged over the past 10 million years". Of the 134,425 species that the International Union for Conservation of Nature (IUCN) have assessed, 37,400 are threatened with extinction and appear on their "Red List". In 2020 alone, the IUCN declared thirty-one more species extinct. In a period characterised by escalating species diminishment, how do we mark these losses?

Extinction occurs within a complex affective economy in which a range of negative emotions are expressed and exchanged: rage, hopelessness, fear, grief, anxiety, desolation, trauma, and loneliness. Grief has emerged as a dominant and powerful emotion in contemporary environmental discourse. The ecocritic Timothy Morton writes that "now is the time for grief to persist, to ring throughout the world" (*Ecology* 185), and Lesley Head has suggested that we need to get better at processing grief because in the Anthropocene "grief is a companion that will increasingly be with us" and for this reason, "it needs to become an explicit part of our politics" (2). How might this grief function not only as a way of marking loss but also as a transformative social and political force for environmental justice? In a world of increasing loss, what role might memorial cultures play in enabling us to grieve for and come to terms with extinction? What are the publics for this grief? What does it tell us about ourselves and our place in this world? It is with these questions in mind that I turn to the work of American artist, Todd McGrain who has stated, "A successful memorial brings something of the past to the present and holds it there for reflection" (6).

Lost Birds

McGrain's *Lost Bird Project* consists of five large-scale statues of birds, each of which has been driven to extinction by humans: the Great Auk, Heath Hen, Carolina Parakeet, Labrador Duck, and Passenger Pigeon. Each of these "memorials to extinction" has been placed in a carefully selected landscape in which the species once lived and thrived in the wild. The bird sculptures are cast in bronze, a hardy if contingent metal that through acquired and applied conditions changes in appearance throughout its life. This metal, used in statuary and memorials since antiquity, places these birds within a tradition of the memorial statues used to mark and commemorate significant human lives or events. Every one of these species became extinct before 1940, which means that they are fading fast from living human memory; these birds are now known through their trace in literature, art, museum displays, and archives. McGrain develops his likenesses by working from photographs and specimens in natural history museums.

I will focus here on his Passenger Pigeon sculpture, arguably the most well-known extinction narrative amongst the birds McGrain commemorates.

When Europeans arrived in North America, Passenger Pigeons were so abundant that they constituted around 40% of the land bird population (McGrain 49). The Passenger Pigeon was a migratory bird with such great numbers that a significant number of narratives about this species focus on how the sky would darken as they flew in huge flocks overhead. For example, Joel Greenberg begins his history of the Passenger Pigeon, *A Feathered River Across the Sky*, with naturalist John James Audubon's recollection of the sun being eclipsed for three days by a flock of birds flying along the Ohio River (1). In the nineteenth century, numbers began to decline dramatically. Due to their sheer abundance and how easily they were caught, Passenger Pigeons had become an efficient and popular source of protein for settlers, and by the mid-nineteenth century, a game industry market had developed. They were also perceived as a threat to agriculture and were actively terminated by farmers through shooting, poisoning, methods of asphyxiation, drowning, and trapping (Greenberg 91–93). They were also shot for sport. By the last decade of the 1800s, the number of wild Passenger Pigeons was estimated to be reduced to a few thousand (Greenberg 159). By 1900, there were three remaining flocks of captive birds and on 1 September 1914, the last known Passenger Pigeon, Martha (named after the First Lady, Martha Washington), died in the Cincinnati Zoo.

Martha was displayed with a plaque stating that she was the last of her kind and died as an "endling". Dolly Jørgensen described the "endling" as the last individual in a lineage, the display of which can "bridge the gap between species extinction as an abstraction and the death of an animal as a concrete event" (134). That Martha was named after the First Lady, gives her a particular place in the national consciousness (her mate was called George) and is testament to this sentiment. Martha's cage, housed in a Japanese-style pagoda, would later become the enclosure for the last remaining Carolina Parakeet, which died in 1918 (Enright 159). Martha's body was prepared for exhibition in the Smithsonian and can now be admired in a 365-degree rotation on their website. From her display in life in an ornate pagoda to her exhibition in death at the Smithsonian, Martha was a spectacle of extinction.

McGrain's sculpture exists in relation to other memorial sites dedicated to this species. It is in part because the Passenger Pigeon has such a documented and interesting history that it has inspired memorialisation. In her essay on monuments to the Passenger Pigeon, Enright contrasts the Passenger Pigeon Monument in Wyalusing State Park with the monument at the Cincinnati Zoo. The Wyalusing monument, erected in 1947 by the Wisconsin Society for Ornithology, is the first public memorial to an extinct animal and marks a distinct point in the memorial culture surrounding animals. This plaque frames the extinction as anthropogenic and reads "...this species became extinct through the avarice and thoughtlessness of man". At the Cincinnati Zoo, the Japanese pagoda which once housed Martha has become the Passenger Pigeon Memorial. To mark the one-hundred-year

Figure 10.1 Passenger Pigeon Memorial, Grange Audubon Center in Columbus, Ohio. Photo Credit: Todd McGrain.

anniversary of the death of the final Passenger Pigeon, this site was redeveloped from a memorial to a single species to an educational resource about extinction and conservation more broadly (Figure 10.1). "While each form of exhibition may include processes of simultaneous remembering and forgetting", Enright writes, "the intentions of the local memorial more explicitly aimed to incorporate the animal and its story into its continually evolving heritage instead of preserving it in an historical past" (166). Jørgensen notes in relation to Passenger Pigeon memorials that they often challenge human exceptionalism. "The oldest passenger pigeon memorials", for example,

> were set up at the same times as monuments to the soldiers of WWII. These birds were also casualties of war, although it was a war on nature, and were understood at the time as worthy as objects of grief. The species was lost, gone from the landscape, and that was worth an expression of sorrow.
>
> ("After None" 196)

McGrain's Passenger Pigeon is placed within the Grange Audubon Center in Columbus, Ohio, a centre dedicated to the conservation of natural environments for birds. This is an environment in which this species would once have been plentiful. McGrain's bird sculptures are all deposited in specific sites which draw our attention to changing and sometimes lost landscapes and which suggest a more significant environmental loss than of an isolated species. While some of McGrain's other birds are accompanied by extensive narrative materials, the Passenger Pigeon is situated in a context dedicated to education and with broader interpretive material already available. The accompanying plaque reads simply "Passenger Pigeon, Driven to extinction 1914". McGrain has noted

> [at the Grange Audubon Centre] they begin every walking and birding tour of the preserve at the Passenger Pigeon Memorial. For me, this is a wonderful way to connect a tragic history with the current efforts to preserve and protect.
>
> (McGrain, private correspondence)

The sculpture, and the story of the demise of the Passenger Pigeon, provides the context for engaging with conservation and bird life more broadly. The place of narrative is important here. In van Dooren's work, he is centrally concerned with the role of storytelling in shaping environmental sentiment around extinction (See 2014). This memorial is a prompt for the public to engage with narrative and to develop greater empathy for other species with whom we have and still share this planet. It also places human impact on the environment and other species at the centre of larger stories about conservation.

The statue itself is a large, sensuous object with a smooth, glossy, curvaceous surface. Significantly, it is cast at human scale (over 6 ft tall), reminiscent of a colonial or commemorative statue of power. McGrain has commented in an interview

> For me, full-size means a size that we can relate to physically. The scale of these sculptures has nothing to do with the size of the bird; it has to do with coming up with a form that we meet as equals. It is too big of a form to possess, but it's not so big as to dominate, the way that some large-scale sculptures can.
>
> (see Gambino)

Its figurative nature and vertical orientation echoes that of monuments from early antiquity to late colonial statues of power (see Doss 37). While this could be read as anthropomorphic, it also offers new patterns of relating to species that were hunted to extinction by (Western) humans. The scale positions the viewer so that prior histories of power and domination are subverted. In coming face-to-face with these creatures, the viewer is invited to stand in a relationship of equals and to cultivate a relation of empathy.

The Passenger Pigeon Memorial can be understood within the context in the United States that Erika Doss describes as memorial mania, "an obsession with issues of memory and history and an urgent desire to express and claim those issues in visible public contexts" (2). As part of this broader nationalist project, memorials can contribute to creating consensus around national narratives and identity. However, as Doss suggests, contemporary memorial mania is "marked by conflict, rupture, and loss and by a recognition among artists and audiences that memorials have the power to stir things up as much as smooth them out" (47). In pointing to human complicity in extinction, McGrain's work stirs up complex feelings about human accountability. In positioning animals as subjects of remembrance and part of the memorial landscape of the United States, McGrain challenges the anthropocentrism of nationalist memorials more broadly. At the same time, these sculptures perpetuate the species-thinking (see Chrulew) that we often see in museums and zoos in which a single animal stands in for its taxonomic type. The smooth stylised silhouette of the statue obscures the detail and specificity of any individual animal. It is not unlike commemorative statues to the unknown soldier in which a single and normatively depicted soldier stands in for all the dead. In this context, the unknown soldier becomes an "everyman", simultaneously abstracted to the national scale and infinitely readable as any individual.

Monuments and memorials can be didactic. This is exemplified by the brochure accompanying replicas of McGrain's sculptures when they were displayed in the Smithsonian Gardens in 2014, to mark the anniversary of Martha's death. It read: "These sculptures compel us to recognize the finality of our loss. They ask us not to forget, and they remind us of our duty to prevent further extinctions" (Smithsonian). McGrain's memorials create a contemplative space in which to acknowledge and mourn prior extinctions and through this reflection come to understand that preventing future extinctions is critically important. His sculptures contribute to the archival evidence that these species once existed so that they remain present (in some form) for future generations. He writes, "The memorials can only point to what is missing, remind us of what we are not seeing, so that we may simultaneously feel the absence and presence of these lost birds" (13). Cast in bronze, which can last a thousand years, these sculptures re-scale time and invite a vaster contemplation of species diminishment. In this way, these bronze birds are dragged beyond temporal constraints and assume the vantage point of immortality from which to bear witness to the ever-escalating extinctions which plague our planet.

Messages from the Edge of Extinction

Chris Jordan's photographic series *Midway: Message from the Gyre* (2009–) depicts the remains of decaying albatrosses taken in situ on the remote islands of Midway Atoll. In 1903, Midway was an important point in the first transpacific cable; in the 1930s, it was an aviation stop-over for Pan American Airways, and in World War Two, it was the location of the 1942 battle of Midway Atoll, which was a turning point for the United States in the Pacific. The island contains an abandoned military base and other infrastructures attesting to this history. Midway has been a National Wildlife Refuge for bird life since 1996 and holds the status of Battle of Midway National Memorial. Today, it is managed by the US Fish and Wildlife Service and is now called Papahānaumokuākea Marine National Monument. Midway is also an important breeding ground for migratory birds, and in particular, for the Laysan and Black-Footed Albatross. Around 70% of the world's population of Laysan Albatrosses make their home on Midway where they have no natural predators. Jordan's photographs are part of a larger environmentally motivated project which also includes his 2017 film *Albatross*, in which he documents the Albatross on Midway Atoll and the impacts of pollutants on their lifecycle. Jordan made this film over the eight years that he visited the region, and it is a poignant account of environmental loss. His images of dead Albatross and the plastics they have imbibed hold a special place in the visual culture of extinction and have been described by Stef Craps and Ida Marie Olsen as "an icon of the Anthropocene" (109).

The Laysan and Black-Footed Albatross fly vast distances over the Pacific before returning to Midway to breed and fledge their young. Mistaking it for food, they ingest the plastics that gather in the region through the atmosphere and waterways. Although Midway is 1,930 km from the nearest significant human population (van Dooren, *Flight* 21), the presence of plastics in their systems is clear and undeniable evidence of the impact of rampant consumerism, even in this remote location and in the vast Pacific Ocean. The birds feed these plastics to their chicks, which means that the Albatross accumulate plastics in their systems throughout the whole life course. The plastics can't pass easily through their bodies and build up until they become fatal. Jordan's photographs depict the remains of birds in states of decomposition revealing their interior cavities and the plastics within them, the detritus of human life – bottle caps, plumbing equipment, fishing wire – all clearly visible in the images (Figure 10.2).

In his book, *Waste*, Brian Thill describes Chris Jordan's images in this series in the following terms: "everything has come spilling out of them like candy from a busted pinãta" (9). These images can arouse feelings of loss,

Figure 10.2 Chris Jordan CF000774 from the series *Midway: Message from the Gyre.*

grief, and guilt, but there is also something profoundly enchanting about these visions of decay. These remains will continue to decay and fade away over time, but the plastics will remain. Michelle Bastian and Thom van Dooren refer to plastics, along with radioactive waste and chemical pollutants, as the "new immortals" because they have "interpellated us into unfathomably vast futures and deep pasts, with their effects promising to circulate through air, water, rock and flesh for untold millions of years" (1). In his book *Anthropocene Poetics,* David Farrier writes about "future fossils", the traces humans will leave on the planet which will become *our* fossil record in the deep future. On the omnipresence of plastics in human lives Jordan has stated:

> And of all the possible pollutants to find in their bellies, plastic is iconic on so many levels as a symbol for our time. For me its defining characteristic is how long it lasts—it is the immortal substance, but in that immortality is the ultimate sterility. And we embrace that sterility, which says something about us; when you look at how we drink plastic-bottled water and wrap our food in it, you can come to see plastic as a desire for protection, an unconscious manifestation of our fear of death. Then

> if you consider what it is made of, petroleum, the disinterred fossil remains of our most distant ancestors, it stands as a symbol on another whole level.
>
> (Jordan in Craps and Olsen 112)

Plastics seem banal. They are familiar, ubiquitous, often inexpensive, and deeply imbricated in modern life. Although plastic is often considered disposable, its temporality is vast: "The sixty-day life-span of a single-use plastic container", Farrier writes, "is informed by [...] the millions of years it took for the raw materials (oil) to form and the ten thousand years it will take for the plastic to break down" (73). Plastics, he goes on to say, are the "projection of a contemporary human presence into the very deep future" (Farrier *Anthropocene* 79). Micro-plastics are less than 5 mm in length, which can result when larger plastics break into smaller and smaller pieces. Because micro-plastics can move easily through our waterways, they are now found everywhere on earth – from our most intimate home environments to the most distant and isolated beaches. "As our trillion pieces of indigestible plastic spread daily across the globe", Brian Thill writes, "they flicker in and out of view, appearing in our storm drains or rolling down the street before being dumped out to sea. We may see them bobbing briefly before they drift away and sink, and are forgotten" (11). Plastics are a good example of what Morton calls a nonlocal hyperobject – something massively distributed in time and space and which challenges our scales of perception. The Anthropocene is the time in which we are becoming more aware of hyperobjects such as climate change, which is vast yet we experience it at a local level (see *Hyperobjects*). Jordan's images bridge the distance between the massively distributed hyperobject of plastic pollution and its concrete impacts. The images are shocking yet also recognisable and familiar. Farrier reminds us that every interaction with plastic "inducts us into a deeply strange and complex entanglement of time scales and processes, as deep pasts and futures hedge each fleeting moment of consumption" (73).

Both the Laysan Albatross and Black-Footed Albatross are classified as "near threatened" on IUCN "Red List" from the last assessment of the species undertaken in 2018. While plastics are a highly visible and publicised danger to these birds, in part through Jordan's images, there are other threats to their survival as a species. The biggest cause of Albatross mortality is becoming bycatch to commercial fishing (see van Dooren *Flight Ways* 29). There is also the likely impact of rising sea levels, which will swallow their existing nesting spots located predominantly on low coral atolls. The Albatross life cycle is a long one. The lifespan of a Laysan Albatross can exceed forty years, and they reach reproductive maturity later in their life, around 7–9 years old. Their clutch size is a single egg and the work of fledging a chick is shared between partners, which are carefully chosen through complex and extended mating rituals. Untimely, Albatross death has a significant impact on the future viability of the species.

In contemplating images of the remains of extinct California Condors, Cary Wolfe asks a series of questions which are instructive for how we engage with Jordan's work: "what do we call these bodies before us? Are they 'corpses', 'remains'? Or just objects, like a rock, a table, or even a leaf? And if not, if they are remains, what are they remains *of*? To whom or to what do they belong? And what, in turn, is owed to them?" (110). He goes on to ask an ethical question: "What shall we do with these remains that are 'delivered over' to us? What will we make of them? And what will *that* make of us?" (114). Jordan's photographs are not of indeterminate dead wildlife but are carefully and respectfully captured as an act of honouring the lives and deaths of these animals. Jordan is aware of the memorial nature of his images, and in the film *Albatross*, this is brought to the surface through the presence of mandalas. In an interview with Craps and Olsen, he comments:

> There is also something about the helplessness of being there after they die, the desire to make some sort of gesture to honor what just happened. So we made mandalas around the birds, using sticks or flowers or rocks or just our footprints in the sand. Usually we just started doing it spontaneously, without any plan or even any words at all. Sometimes it was a circle shape, or other times in the shape of an egg around the bird, maybe like the egg it came from, a symbolic return to the mother. We made a few hundred or so of those, maybe some of them are still there.
>
> (Jordan in Craps and Olsen 126–127)

Jordan's images show birds that are laid out for mourning. He is very clear that he does not manipulate the bird remains but photographs them respectfully and as he finds them. As images, each photograph is unique and honours the death of a single animal. However, they are also part of a series, and as a collective, they tell a larger story about human planetary impact. These are also memorials to the living, to a species that is threatened but not yet extinct. Thom van Dooren writes of the "dull edge" of extinction that occurs not through a single event, but as a "prolonged and ongoing process of change and loss that occurs across multiple registers and in multiple forms both long before and well after this 'final' death" (*Flight Ways* 58). The Albatross of Midway shows us this dull edge, the many untimely deaths that build toward the finality of extinction. Returning to Wolfe's question: what is owed to these birds? Jordan's photographs remind us that the deaths of these birds are anthropogenic in nature, but like McGrain's memorials, the intention behind these artworks is to elicit social and political change. The accessibility of Jordan's work and its capacity to stimulate public sentiment is important to the artist who has made his film *Albatross* freely available through the internet and without a distributor so that it can, according to the website, be "spread around the world". Similarly, the *Midway, Messages from the Gyre* series is reproduced on his website for public accessibility. This is a decision that refuses forms of neo-liberal individualism and consumerism that endangers these birds in the first place. As a public, we have things to consider: How might we turn grief and mourning into environmental action? What responsibility do we carry for the detritus of our lives? How is this responsibility distributed racially, socio-economically, geographically? What can we do to ameliorate future extinctions? What can we do to mark those species we have lost?

Both McGrain and Jordan invite us to grieve animal life and through this they challenge human exceptionalism and create a contemplative pause in which to think about human environmental impact and responsibility. These artworks ask us to think about individual and distinctive animals and species, but, ultimately, they invite us to consider our complicity with and responsibility for the sixth mass extinction event that is currently underway. They also invite us to think about our value systems and the place of animal life within them. If the Anthropocene is the age of human impact, it must also be the age of human responsibility. However, despite a renewed focus on human impact and responsibility, the escalating rate of anthropogenic extinctions requires that we more stringently critique the anthropocentrism in our systems of meaning and value.

Conclusion

John Charles Ryan frames the Anthropocene in relation to extinction as a time of "profound loss and also of profound loneliness" (190) in which we must find ways to grieve more-than-human lives. The eco-grief attending extinction is often framed individually and psychologically as eco- and climate anxiety (Pihkala), climate trauma and pre-traumatic stress disorder (Craps "Climate Trauma"), Anthropocene disorder (Clark), and solastalgia (Albrecht). While not denying the very real psychological distress experienced by individuals in the face of environmental crisis, memorial culture points to the way that emotions are collective and shared. This means that eco-grief also needs to be processed as a cultural and political phenomenon with a public life. To memorialise something is to give it cultural value but it is also marks out time and space for mourning.

In her work on "grievable lives" Butler writes of the benefits of "tarrying with grief" (23). She asks:

> If we stay with the sense of loss, are we left feeling only passive and powerless, as some fear? Or are we, rather, returned to a sense of human vulnerability, to our collective responsibility for the physical lives of one another? The attempt to foreclose that vulnerability, to banish it, to make ourselves secure at the expense of every other human consideration, is surely also to eradicate one of the most important resources for which we must take our bearings and find our way.
>
> (23)

While Butler is specifically engaged with the question of the human and of human lives, this work provides us with a framework for thinking about the place of humans in a larger network of vulnerable lives and it frames recognition of this vulnerability as a resource for change. Extending grievability beyond the human is ethically and politically urgent. As Stef Craps suggests, grieving for extinct species can "galvanise us to take positive action on their behalf" ("Ecological Grief"). But it also extends the very notion of political community and sociality. What does it mean to extend the category of "grievable lives" beyond the human and to creatures that no longer exist? Taking Butler's notion of grievability out of her anthropocentric frame, grieving extinct species can be seen as a political act. "Many people think of grief as privatising, that it returns us to a solitary situation", Butler writes, "but I think it exposes the constitutive sociality of the self, a basis for thinking a political community of a complex order" (19). In Butler's work, she discusses the way emotions such as grief transform us through rendering us ecstatic or beside ourselves. She describes how "passion and grief and rage [...] tear us from ourselves, bind us to others, transport us, undo us, and implicate us in lives that are not our own" (20). To be beside oneself is, for Butler, a form of dispossession and a fundamental condition of living in the world with others. When deployed in relation to extinct animals or species, grief reminds us that we are grappling with extinction in a multi-species world. In his case study of the Hawaiian Crow, van Dooren picks up on the way that grief, particularly collective grief with other species, reminds us that we live and die in a world we share with nonhuman animals. He questions what it is that is lost when a species becomes extinct, and concludes that this loss goes well beyond a discrete species or a position within an ecosystem. "Far more than 'biodiversity' in any narrow sense", he writes, "mourning crows reminds us that whole modes of life, whole ways of living and dying in company with others, are disappearing – nonhuman languages, socialities, perhaps even cultures" (137). How do we mourn for what has been lost when that loss is not just a species but a way of life, and the networks and entanglements, including cultural, that constitute that species' existence?

In examining the ways in which species loss is mourned, my intention is to offer a more diverse and affecting vision of the commemorative landscape that picks up on Butler's call for a political community formed in and through extinction grief. It is notable that the species memorialised here are charismatic and have been caught up in human lives in both literal and symbolic ways. In being more relatable and proximate to humans, they are easier to include within the remit of "grievable lives". Questions remain about how we meaningfully memorialise the species loss occurring beneath our usual scales of perception and beyond our perceived relational sphere – the extinctions that occur without us even noticing them.

We need to take negative affects seriously and to give it spaces in which to flourish. Memorial culture has an important role to play in facilitating public grief in the Anthropocene. Rather than focussing on the individual eco-grief and trauma, thinking about how emotions circulate and are exchanged can enable us to start to understand how public grief and mourning might act as potent catalysts for environmental action. As van Dooren and Rose remind us the "reality is that there *is* no avoiding the necessity of the difficult cultural work of reflection and mourning. This work is not opposed to practical action, rather it is the foundation of any sustainable and informed response" (377). Shifting to a framework in which we acknowledge the ways that environmental grief is a shared project of public mourning holds out hope for the dynamism and possibility within the public cultures formed in and through the extinction crisis.

Note

1 Thank you to Toby Juliff and Keely Jobe for their thorough and incisive criticism which has no doubt made this chapter stronger.

Works Cited

Albrecht, Glenn. " 'Solastalgia': A New Concept in Health and Identity." *PAN: Philosophy, Activism, Nature* 3 (2005): 41–55.

Bastian, Michelle and Thom van Dooren. "The New Immortals: Immortality and Infinitude in the Anthropocene." *Environmental Philosophy* 14.1 (2017): 1–9.

Butler, Judith. *Undoing Gender*. Routledge, London, 2004.

Chrulew, Matthew. "Managing Love and Death at the Zoo: The Biopolitics of Endangered Species Preservation." *Australian Humanities Review* 50 (2011): 137–157.

Clark, Timothy. *Ecocriticism on the Edge: The Anthropocene as a Threshold Concept*. Bloomsbury, London, 2015.

Craps, Stef. "Climate Trauma." *Routledge Companion to Literature and Trauma*. Colin Davis and Hanna Meretoja (Eds.) Routledge, London, 2020, pp. 275–284.

Craps, Stef. "Ecological Grief." *American Imago* 7.1 (Spring 2020): 3.

Craps, Stef and Ida Marie Olsen. "Grief as a Doorway to Love: An Interview with Chris Jordan." *American Imago* 7.1 (Spring 2020): 109–135.

Enright, Kelly. "Exhibiting Extinction: Martha and the Monument, Two Modes of Remembering Nature." *Cultural Studies Review* 25.1 (September 2019): 154–171.

Farrier, David. *Anthropocene Poetics: Deep Time, Sacrifice Zones and Extinction*. University of Minnesota Press, Minneapolis, 2019.

Gambino, Megan. "Bronze Sculptures of Five Extinct Birds Land in Smithsonian Gardens." *Smithsonian Magazine*, Accessed April 4 2014. https://www.smithsonianmag.com/smithsonian-institution/bronze-sculptures-of-five-extinct-birds-land-in-Smithsonian-Gardens-180950366/

Greenberg, Joel. *A Feathered River Across the Sky: The Passenger Pigeon's Flight to Extinction*. Bloomsbury, London, 2014.

Head, Lesley. *Hope and Grief in the Anthropocene: Re-conceptualising Human-Nature Relations*. Routledge, London, 2016.

Intergovernmental Science-Policy Platform on Biodiversity and Ecosystem Services (IPBES). *Global Assessment Report on Biodiversity and Ecosystem Services*, 2019. https://ipbes.net/global-assessment

Jørgensen, Dolly. "After None: Memorialising Animal Species Extinction Through Monuments." *Animal Count: How Population Size Matters in Animal-Human Relations*. Nancy Cushing and Jodi Frawley (Eds). Routledge, 2018, pp. 183–199.

Jørgensen, Dolly. "Endling, the Power of the Last in an Extinction-Prone World." *Environmental Philosophy* 14.1 (2017): 119–138.

McGrain, Todd. *Email to Hannah Stark*, 25 May 2022.

McGrain, Todd. *The Lost Bird Project*. University Press of New England, 2014.

Morton, Timothy. *Ecology Without Nature*: *Rethinking Environmental Aesthetics*. Harvard University Press, Cambridge, MA, 2007.

Morton, Timothy. *Hyperobjects: Philosophy and Ecology After the End of the World*. Minnesota University Press, Minneapolis, MN, 2013.

Panu, Pihkala. "Anxiety and the Ecological Crisis: An Analysis of Eco-Anxiety and Climate Anxiety." *Sustainability* 12.19 (2020): 7836.

Red List for Threatened Species. Accessed 27 October 2022. https://www.iucnredlist.org/

Ryan, John Charles. "Kelp." *Living with the Anthropocene*. Cameron Muir, Kirsten Wehner and Jenny Newell (Eds.) NewSouth Publishing, Montgomery, AL, 2020, p. 190.

Schuster, Joshua. "Life After Extinction." *Parrhesia* 27 (2017): 88–115.

Smithsonian Libraries. "The Lost Bird Project: Todd McGrain Sculptures." Accessed 27 October 2022. https://library.si.edu/sites/default/files/pdf/general_pages/lost-bird-project-smithsonian-gardens.pdf

Thill, Brian. *Waste*. Bloomsbury, London, 2015.

van Dooren, Thom. *Flight Ways: Life and Loss at the Edge of Extinction*. Columbia University Press, New York, 2014.

van Dooren, Thom and Deborah Bird Rose. "Keeping Faith with the Dead: Mourning and De-extinction." *Australian Zoologist* 38.3 (2017): 375–378.

11 Psittacine Extinction Story

I Once Loved a Lorikeet

Hélène Frichot

On quiet streets in the early days of the pandemic, and during the small window of time allowed for exercise out of doors, neighbours would remark to each other in wonder about the birdsong. Had it ever been so vivid, so distinct, so hopeful, even? People, who had never given the time of day to these winged environmental co-habitants, craned their necks – looking towards the sky, nearby powerlines, tree branches – and a soft expression would fall across their faces. This was during the first lockdown in *naarm* (Melbourne, Australia), when all but essential workers found themselves suddenly cloistered in domestic surrounds. It was when lockdown life was still a novelty, and people mended, baked, checked in on elderly neighbours with small handwritten notes, and the city quietened down for once, a curiously peaceful hiatus only darkened by the lack of available knowledge on the swiftly circulating novel coronavirus and what was in stall for us next. By the time the sixth lockdown was imposed on exhausted Melbournians, the uncanny thought of inhabiting a sixth lockdown amid a sixth great extinction event had begun to trouble me. At seventy-eight days duration, the sixth lockdown was the second longest, the longest having been the second lockdown, at one hundred and eleven days. By the time Melbournians exited their sixth lockdown, they had been under conditions approximating house arrest for two hundred and sixty-three days, purportedly making our experience of lockdown the longest in the world (Miller). As for the sixth great extinction event, though its aftermath is impossible to imagine, its anthropogenic causes are clear to see.

When I look back at these six lockdown events, I can't quite grasp what we went through, nor how we endured. What trauma to the collective psyche have we been left to cope with? Trauma, as Catherine Malabou argues, changes us. We become not simply other than what we were but, at the limit, "Wholly Other," to the point of no longer recognising ourselves (3). Malabou offers literary examples of the "destructive plasticity" wrought by the radical contingency of a traumatic event, what she calls the "unexpected arrival of catastrophe" (90). Malabou turns to literary examples to illustrate the psychic break that trauma enacts. There is K. in Kafka's *The Metamorphosis* who wakes up one morning as an insect, finding himself radically

DOI: 10.4324/9781003315957-12

unfamiliar to himself, even as he holds some vestiges of his former incarnation in his recognition of family relations (14–15). There is the famous scene from Marcel Proust's *In Search of Lost Time*, where former acquaintances are rendered unrecognisable with age (50–52). Less than a gradual process of becoming-old, they are suddenly rendered old, as though time had been folded, abruptly collapsing in on itself. Malabou effectively draws attention to the formal qualities of destructive plasticity, understood as an event that imposes a new form on an old one (6). Things and their relations take on a different shape. Could this destruction of a former identity, through war, plague, and even anodyne and daily forms of violence, result in the production of other kinds of memorial forms? In her discussion of memorial culture, Adrian Parr forges a direct association with trauma and the way it "gnaws its way through the social field" (1). Trauma is disruptive and contradictory, and demands some form of material response, which often resolves itself into a memorial, even if the memorial is embodied only so that we can eventually shake it off our collective traumatised body in order to forget. With respect to the grief and trauma aroused by these pandemic times, it's not over yet. Considering concatenating environmental crises, no doubt our troubles have only just begun.

I am further convinced that these near unprecedented disruptions to the environment and everyday experience have produced a deranging effect on my memory and sense of time passing. A collective trauma was being experienced during extended lockdown and continues to be experienced. Layers of the past have been shuffled, and peaks of the present have suffered parallax rearrangements. Subjective points of view have been shattered and rearranged. This general derangement eventually led me to ask myself questions such as: where does that memory affected by trauma or disruption reside inside of me? Does it live in a part of my body, like my big toe, or my left arm? What is it that happens when a memory suddenly looms up, as though out of the depths, and overwhelms me, returning me to its moment in vivid detail, only to dissipate again? Is it the same memory, or rather a process of fabulation, a construction I continually work upon for as long as I endure? Then, because the theme of death has been ever present in these pandemic times: Where do my memories – so vivid, so real to me, so there, so apt to be repackaged as stories, given shape with a line, form, a chunk of marble – where do they go when I die? Do the effects I leave behind when I die embody personal memorials or memorabilia?

Returning to bird song, its presence, and its absence, I couldn't help but be reminded of the haunting opening parable from Rachel Carson's *Silent Spring*:

> There was a strange stillness. The birds, for example – where had they gone? Many people spoke of them, puzzled and disturbed. The feeding stations in the backyards were deserted. The few birds seen anywhere were moribund; they trembled violently and could not fly. It was spring without voices.
>
> (2)

The birds in my parable are notable in that their voices are suddenly heard as the everyday noises of air, road, and rail traffic, as well as urban industry and construction projects, have been quietened. Yet, the sudden clamour of birds enunciates a presentiment of increasingly uncertain environmental futures. Unprecedented events are just waiting to happen.

Memory leads inexorably to the desire to memorialise, to grasp that which threatens to slip away, which will, in any case, slip away. Memories understood as embodied phenomena are preserved for only as long as the material allows, and in any case are reliably unreliable. Memories often wander hand in hand with grief and trauma. There is a cut. Someone, or something, is lost to us. They were there, only just then, and now they are gone. We attempt to cling to their memory. With their passing, the world has changed irrevocably. "Stop all the clocks, cut off the telephone" (Auden). To transmogrify the stages of grief into something concrete, to be held, visited, returned to, worn away with visitations, results in the formal manifestation of a memorial and the development of memorial culture. Where the memorial stubbornly persists, memories shift and change, loom up vivid, only to fade away. Imagine returning to a memorial that improbably changes shape on each visit, it's not how things work. Instead, there is something of a disjuncture between the memorial and the memories it hopes to secure in place. Gilles Deleuze and Félix Guattari counter habitual assumptions of memorial culture when they discuss their concept of the monument, arguing that: "The monument's action is not memory, but fabulation" (*What* 168). What holds the memory together is whatever construction endures, whatever has enough structural strength to support a compound of sensations, whether that be an architectural body, a human body, another kind of body altogether, and why not, an essay as experimental test site and body. The memorial for Deleuze and Guattari is less about the past than about future encounters and what happens from the midst of things as they unfurl.

Considering grief and the memorial's response to this affect, what happens when this intensity of feeling is not directed at the loss of a fellow human being, but towards a member of another species, a non-human being? Responding to the challenge of this edited collection dedicated to how we might reckon with species loss amid the sixth great extinction event of this our Anthropocene era, and how a thinking with memorial culture might avail us in this process, leads me to reflect on a personal encounter I recently experienced of the cross-species kind. And so, this essay is dedicated to a Rainbow Lorikeet I once loved, who entered my life between the second and third pandemic lockdown in *naarm* and spent some fourteen weeks with my family before departing abruptly one day. I should hasten to add that the Rainbow Lorikeet is not a species threatened with extinction, and in parts of New Zealand and Australia is even considered a pest. I surprise myself nonetheless when I confess that I still feel the loss of this bird, who was certainly no pet, rather a visitor with whom I briefly entered a cross-species relationship. What I feel is grief, and as I've suggested, this is an affect that is hooked up with the compulsion to memorialise. More than that, my relationship with Lori has intimately alerted me to the seething ecological relations

Figure 11.1 Rainbow Lorikeets in Birch Tree. Carlton North, Melbourne. Author photograph.

with which I am bound, and so, more than grief, this is what many are now calling "eco-grief" (Engstrom; Pihkala; Hron). I have no idea whether Lori is meanwhile alive or dead, simply not here. I continue to feel Lori's absence as though it were some part of my embodied self that had been detached and floated wide. I am no longer able to remain indifferent (Figure 11.1).

There is an exquisite sequence of moves by which Deleuze and Guattari describe the material contraction of memories, or, rather, their preferred term, "fabulation": *vibration, clinch, withdrawal.* The vibration of a simple sensation is followed by the clinching together or embrace of two sensations, and then, the inevitable withdrawal, when two sensations draw apart even though they remain connected through the communicating medium of the air or the light. Reading Deleuze and Guattari, Parr identifies these as the three syntheses of desire, which organise what she calls memorial culture: connective, conjunctive, disjunctive. Following the encounter with the intensity of the event, the work of memory can follow these syntheses and arrive at a reformulated even utopian societal composition, or else, society can become more paranoid and fascistic as memory comes to be commodified and thereby stultified. In either case, memorial culture "organises a social habitat" according to Parr (5–6; Frichot "Connexa"). We might be habituated to individuating our sorrows, but at the scale of a body politic, memorials aim to speak for the collective social subjectivity. We are reminded here that memory is not only a personal affair, but leaks and contaminates. Memorial culture when expressed as a positive and affirmative force, and not a passive and negative one, moves away from memory, per se, towards what Parr calls "memory thinking" (12) as a process of construction.

In the three Anthropos-scenes that follow, I first introduce a member of the Psittacine family, Lori the Rainbow Lorikeet. Lori's impact on my cross-species reflections on extinction and memorial culture cannot be underestimated. Cycling or walking around the neighbourhood I still wonder whether I am sharing the same air as my more-than-human friend. When a pair of Rainbow Lorikeets passes, and one calls out and dips a wing, I wonder whether it is my dear friend dropping by. In venturing the connection between the sixth great extinction event and memorial culture, I'll look to Lori the Rainbow Lorikeet and our encounter and brief friendship, for advice. Even though I cannot know whether my feelings were truly reciprocated, Lori will act as my guide.

While the Rainbow Lorikeet is considered a pest in some parts, the conservation status of its cousin, the Orange Bellied Parrot, is identified as critically endangered. In scene two, I introduce the Orange Bellied Parrot (OBP), one of the rarest birds in the wild and one of only three members of the psittacine family who migrate to breed. Placing a sometime pest next to an endangered family member draws attention to the vulnerable ecologies we inhabit, located as we are amid cross-species entanglements. Should the efforts of a memorial culture acknowledge not only extinct and endangered creatures, but also what Anna Tsing and her collaborators call feral ecologies (Tsing, Swanson, Gan, Bubandt; Tsing *Mushroom*)? A feral ecology is characterised by the flourishing of one species at the cost of others, jellyfish in the Black Sea, for instance, that destroy fish and fisheries, emptying the sea of other creatures while enjoying their own accentuated capacity for life (Tsing et al.). While the extinction of one species can be the result of the excessive flourishing of another species, the point Tsing and her co-researchers want to make is that such expressions of excess are usually aided and abetted by human industrial activity. You see, we are the monster who has created such ecological imbalance, while at the same time becoming the creature most anxious to preserve life at the brink of extinction. Memorial culture locates value in those species at greatest risk of exiting the world-historical scene, the irony being that it is increasingly the human species-being that has accelerated this exit. A consideration of the OBP can be located at the threshold of extinction, where the early moves of a memorial culture are expressed as efforts in research and conservation. The concerned researchers and other communities who gather around such efforts might be understood as performing their own rituals of preservation by attempting to hold on to something in anticipation of its imminent demise.

The involvement of the institution of the zoo in such endeavours speaks to the transformation of a colonial institution of collection and display into one of research and conservation, including the affective arousal of a concerned or simply curious public. In scene three, I raise questions about the transformation of the institution of the zoo, observing how its remit has increasingly been extended to breeding programs and rescue efforts aimed at staving off extinction. There is the risk of the spectacular and carnivalesque creeping into the chance to take what might be a last look at a living specimen of a near extinct species: Could this be the promise and threat of extinction on display, life memorialised in advance of its inevitable passing away? The zoo is closely connected to those proud and "civilising" colonial institutions dedicated to world expositions and the display of edifying, strange, and wondrous artefacts, animals, and even humans from "exotic" lands. It has historically sought to demonstrate clear and taxonomical divisions. In its interface with a public domain, the zoo of the nineteenth century, as Sarah Amato argues, explored the boundaries between animals and humans, led to reflections on socially acceptable behaviour, and participated in the formation of a rapidly expanding consumer culture (5).

To collect, to preserve, to display for science and pleasure, these are the dubious means by which we humans have sought dominion in the Animal Kingdom, as well as the dominion of coloniser over colonised, purported civiliser over barbarian, with all the ramifying environmental devastations that we now know follow. Lori helps me ask: For whom, finally, are conservation efforts being undertaken? Who or what did I think I was saving when I recovered Lori as a chick under a local tree? When we task a memorial culture with the preservation of the memory of a species rapidly vanishing from our ranks, what is it that we are hoping to achieve? Will we arrive at a hoped-for environmental utopia, or will the reorganisation of the social habitat prove insufficient? Is the environment, after all, indifferent to all human efforts? Is it, as philosopher Isabelle Stengers claims, that Gaia "*asks nothing of us*" even as she devastatingly intrudes on our daily affairs (46)?

A Rainbow Lorikeet I Once Loved

In the early days of December 2020, when it seemed that, after two lockdowns, we might finally be in the clear, I was walking Dog down the wide, grassy, median strip that leads into town. The grassy median strip outside our rented townhouse had been one of those inner-city stretches of grass crowded with pandemic picnics in small clusters on sunny days (size of cluster depending on the current lockdown rules determined by the state government of Victoria). Dog had picked up a scent and pushed his nose eagerly towards the base of a Canary Island Palm. Just before he got his teeth around it, I managed to scoop up a small, fuzzy, grey quivering form: a baby bird! What next? A project! I enrolled my younger of two sons, and together, we sourced a shoe box and composed a makeshift nest: every suburban family's response in an abandoned or sick bird scenario. I discussed with my son how the small bird's survival was unlikely, but that we could do what was possible to look after it, and who knows, maybe it would learn to fly away and re-join its mob. Next step was a call to the local wildlife emergency hotline who informed me that the public are not to take indigenous wildlife into their care. There was the threat, furthermore, of the dreaded and highly infectious beak and feather disease. I was to take the fledgling to the nearest animal hospital, or vet, but it sounded very much like their future was grim for it was likely to be euthanised. So, I resisted all advice, and took a deep dive into "desktop research" (the land of Google) to work out how best to support a baby bird without access to their parents.

The species identification was relatively easy. The faintly blue head, distinct hooked beak, and local bird populations gave it away as a Rainbow Lorikeet. Later, I would read of how the introduced Myna bird was in the habit of kicking local fledglings out of their nests and reclaiming the nests as their own. In my imaginary, the Myna, one of the most destructive introduced species in my local habitat, was the culprit. But how did such a culprit arrive if not in the hulls and stowaway nooks and crannies accompanying human patterns of migration and colonisation?

So began the intensive labour of attending to a baby Rainbow Lorikeet. At estimated age five weeks, Lori (named by my son) needed to be fed every 3 to 4 hours via a syringe (I did not have a special feeding spoon) with a soft paste heated to just the right temperature so their gut could manage. All equipment had to be carefully sterilised before and after use. The first night, I happened to have some mashed sweet potato ready-to-hand, then we added some powdered rice baby food, then we finally bought some dedicated baby Lorikeet formula: just add boiled water. Lori gobbled it up and pooed it out the other end, careful to shit as far from the centre of their makeshift nest as possible. I never could tell whether Lori was male or female, only a size comparison, and failing that, genetic testing enables a determination of sex.

After a week, the shoebox was replaced with an easier to clean homemade breeding box (hygiene is crucial) and a make-shift red heating light was set up. The brooding temperature is likewise an important environmental factor in support of a nestling. As Lori grew from day to day, and then week to week, their feathers sprouted out of wonderfully coloured feather buds, the pinfeathers shedding their translucent shafts of keratin and unfurling. Lori's character evolved, and it took a special shine to their main feeder, me, and accomplice, my son. Fortunately, as Lori grew, the hours between feeding times likewise lengthened.

As our familial relationship developed, so did my concern for Lori's future. Would Lori continue to thrive? Would Lori learn to fly? Would Lori wean themselves from my hand-feeding regime? Would Lori be able to live in the inner-city suburban "wilds" once released? Would Lori turn out to have the dreaded beak and feather disease, and then what? A relationship was taking shape in the comfort of my own domestic environment, where as a family we had found ourselves in and out of captivity, suffering the closing in of familiar walls when our lockdown regulations had become most claustrophobic and intense. Lori had arrived in our home during a promising window of time between lockdowns, when it seemed that the world was opening up again. There was no mistaking Lori's growing attachment to me, and this filled me with a mixture of joy and dread. Lori turned out to be a clown and a great humourist and I would deeply miss their company, once gone. Yet Lori was no pet. I didn't believe in keeping domesticated, caged birds. I was determined Lori would regroup with their mob. At the same time, I began to suffer an acute melancholy, for Lori had clenched their sharp claws deep into my heart and their beady eyes knew me. What kind of environmental subjectivity was I becoming through my intimate connection with Lori? At the time it had not occurred to me that the one thing our cross-species relationship had in common was a shared experience of being incarcerated for our own good and even for the health and well-being of our respective species. For a suite of weeks, I became Bird Lady, that crazy lady.

The Orange Bellied Parrot

Lori, as I've explained, was not a member of the psittacine family threatened with extinction, but a noisy and messy bird some regions consider a pest. I first came across the plight of the critically endangered Orange Bellied Parrot, Lori's relative, while undertaking a collaborative and instructional art project in 2009. At the invitation of the former director of what was once the Centre for Ideas at the VCA (Victorian College of the Arts) Elizabeth Presa, I was engaged in a local episode of Hans Ulrich Obrist's distributed instructional art project called *Do It*. Taking on the allocated role of philosopher, I wrote ten instructions as an invitation to a group of art students Elizabeth was working with. Instruction number three reads:

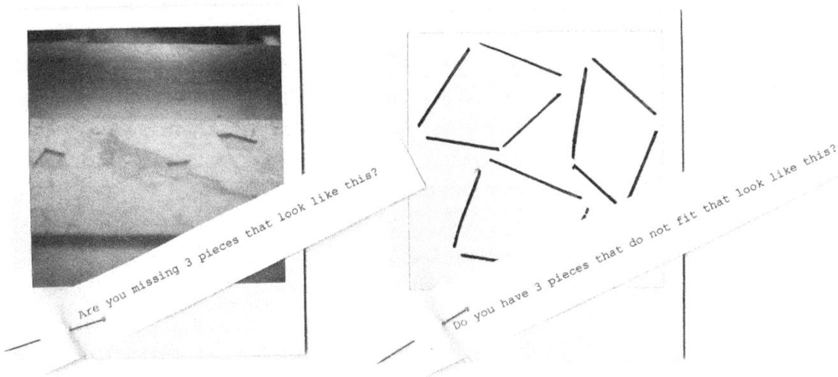

Figure 11.2 Orange Bellied Parrot jigsaw puzzle with missing pieces. Photo credit: Anna Miller Yeaman.

3. Take Three jigsaw puzzle carpenters

With two people you feel ambivalent about build a series of jigsaw puzzles that do not work.

The pieces should not fit right and parts should be missing.

The jigsaws should be made from 0.5 plywood (not cardboard).

The images on the jigsaw puzzle should be composed of strange hybrids of endangered Australian flora and fauna.

Document, then send to my two young boys for testing (puzzles will be returned after testing).

Address: …

One of the young participating artists responded and posted to my home address a jigsaw puzzle she had constructed depicting an Orange Bellied Parrot. She also sent along a stuffed toy version for my sons. We worked out the jigsaw puzzle together, documented it, and then returned it to the artist along with a short extinction story. The stuffed animals have meanwhile travelled the world with us, from Australia to Sweden, then Germany and back down under again, emerging as a renewed surprise every time we unpack a new home. During their travels, some of the orange and green felt feathers on the stuffed parrot have begun to moult. I never met the artist, though recently corresponded with her briefly, only to discover she had moved out of the city and into the country where she tells me she has many more opportunities for cross-species encounters (Figure 11.2).

In 2008, at much the same time as this participatory art adventure, the bird man Sean Dooley was writing urgently of the stricken habitat of the

OBP, whose Victorian feeding grounds had become increasingly inhospitable due to habitat loss and invasive predators:

> All along the west coast of Port Phillip Bay, the OBP's former habitat has been replaced with anything from housing developments to armaments compounds. The few remaining patches of salt marsh are being overrun by invasive weeds. Thick with feral predators, trashed by illegal recreational vehicles and horse riders, and after ten years of below-average rainfall, they have failed to produce the kind of prolific seeding that keeps a parrot going through a long, hard winter.
>
> (Dooley)

One of its habitats was at Fishermans Bend along the *Birrarung* (Melbourne's Yarra River). Once a salt marsh, Fishermans Bend is an industrial wasteland awaiting development. And today, fourteen years after Dooley's article was written, still awaiting development amid stalled plans for a research facility bringing together institutions of higher education and industry on a site which, with rising sea levels, may well be under water in just a few more decades. These are feral ecologies, transformed through human activity, places where the entanglements of human and more-than-humans are at play creating not just local but planetary effects.

In a recent episode of ABC's (Australian Broadcasting Commission) *Catalyst*, Dr Ann Jones and BirdLife's Sean Dooley visit a sewage farm, the Western Treatment Plant, that has transformed a local habitat into an unlikely wildlife sanctuary (2022). It is now an internationally recognised and protected wetland. Dooley explains that the high levels of nitrate and phosphate provide food for the shorebirds, some of whom have flown as far as 12,000 km from their breeding grounds. Suddenly, Jones and Dooley spot an OBP in the wild, a highly unlikely event. Their voices quaver with excitement. First, they hear its call – much of bird watching is apparently about listening, waiting to hear a call – and then, "yes, yes, yes...," they spot two birds through their respective binoculars, and then, a third. It's extraordinary, their excitement is visceral. A leg band is visible on the birds that have been spotted, indicating their involvement in the human project of conservation. With the concerted efforts of conservationists and researchers, as well as associated institutions such as the zoo, the mere fifty OBPs in the wild have increased to sixty-two. Jones exclaims that the fact it is right here, on the edge of one of our biggest cities in Australia, is "mind blowing!" Dooley observes that "we thought they were functionally extinct" and it has been five years since he has spotted one in the "wild." Here remembering that the "wild" in question is also a sprawling sewage farm, a complex feral scene of cross-species interactions can be assumed. Each keen birdwatcher admits to their emotions, and the great significance of the encounter.

A current champion of the OBP is Dr Dejan Stojanovic, a researcher at the Australian National University (ANU) who is based in Tasmania with ease of access to the breeding sites of the OBP.[1] He explains that while breeding programs are making a positive change, they are not enough to save the species, which equally depends on the reparation of habitats, as well as burning regimes to render available the seeds and nuts necessary to the OBP diet and means of managing predators, such as the sugar glider. Unlike the Rainbow Lorikeet whose specially brushed tongue is designed for extracting nectar from flowering natives, the OBP feeds on seeds. A lack of females in the wild, outbreaks of Beak and Feather disease where OBPs breed, habitat loss, the increase of predators, and noxious weeds all contribute to their small numbers. Even when Dejan explains to me that the numbers at the last count have jumped to two hundred in the wild, he warns that the species remains precariously balanced at the threshold of extinction. Strong headwinds in the Bass Strait as they make their journey between breeding grounds and feeding grounds can knock off large numbers. This is a stubborn bird who makes the difficult migratory journey across the Bass Strait between the mainland and south-west Tasmania twice a year, across to the mainland in Autumn to feed, then back to Tasmania to breed across the Summer.

I'm interested in the environmental affects aroused by the cause, the collective dedication to the OBP. A community of concerned researchers is forming around the OBP. I'm convinced that I read expressions of joy on the faces of the research group when I survey the photographic documentation found on the Difficult Birds website, where visitors are invited to support the cause with donations.[2] There might be a collective joy aroused in spending time in the field, but Dejan admits that their research publications can feel like the description of an unfolding catastrophe. A dark, gallows humour is often shared by the group as a means of coping. They are a research community brought together in their shared love of the OBP, and what they are collectively doing is creating a life support system. In being a collective body they are, I can only assume, also marked by trauma, and thus plastically transformed in their nature-culture encounters.

I am told that modelling demonstrates that without the measures currently being taken, crucially combining habitat intervention with captive breeding programs supported by zoos or sanctuaries such as Healesville Sanctuary, Moonlit Sanctuary, and a state government facility in Tasmania, the OBPs would be extinct within five years. Raising public consciousness is important, and this is where information pages on zoo and sanctuary websites further contribute, but it still comes down to funding, of which there is simply not enough, especially given the wealth of a nation-state such as Australia. Why should we bother? I dare to ask. To which Dejan responds that if we cannot care for the OBP, a popular and attractive species, then what hope have we of conserving other species? We compare parrots and slugs, for instance, and he explains how images of the Tasmanian Masked Owl receive far fewer likes on Instagram than the OBP. What can be discovered here is the restricted extent of human sympathy, happy to clamour about a beautiful bird like the OBP, while remaining disinterested in the rest. The challenge to an ethics of care is at stake here, likes and dislikes need to be placed to the side, and broader systemic environmental stakes considered.

What did I learn during the period of close contact with my cross-species other, Lori the Rainbow Lorikeet? The Rainbow Lorikeet has a distinctive brushed tongue for gathering nectar, and their food passes down their oesophagus through what's called the crop, which means that close attention must be paid to the swelling of the crop during feeding, to avoid over feeding. Graduating from their fledgling formula, they need access to flowering natives amid which they can forage. They nest in tree hollows, of which I have found many on my neighbourhood walks, and they patrol in pairs, as well as congregate in larger flocks. Listen to their call and return across the pale skies of Melbourne, as they check in with each other while patrolling the heavens. What I learnt was an intense feeling of cross-species care.

A Visit to the Zoo

Australian zoos collaborate with research groups and government bodies, specifically the Government Department of Environment, Land, Water & Planning (DELWP), on breeding programs dedicated to the OBP. The Melbourne Zoo; the National Zoo and Aquarium, Canberra; Zoos South Australia; Moonlit Sanctuary, amongst others, share with the concerned public information about the critically endangered status of the OBP and other species. These are issues of public interest; this is, after all, a world under duress that we share across species, a world we humans have helped to produce. Even if the zoo appears to place a taxonomical dividing line between us and them, between the political animal that the human is supposed to be – at least according to Aristotle – and those other species closer and further away from us on the great tree of life, the zoo is ever at risk of merely reasserting human exceptionalism as a norm. As Ralph R. Acampora plainly puts it: "Throughout its past the zoo has demonstrated a relational dynamic of mastery" (1). A visit to the zoo means surveying and bearing witness to a taxonomically ordered multi-species captivity machine. We are captivated by the spectacle secured through the captivation of animals that are not us. The proximity of the endangered tiger to the glass, the polar bear found so far from the melting polar cap: an artificially manufactured sense of intimacy is procured with each captured animal only to better secure our human distance and distinction (Figure 11.3).

The situation of zoos in the age of the Anthropocene and how the zoo, formerly a colonial institution, comes to be resituated as an institution dedicated to conservation is the kind of question researchers such as Gail Pini, Matthew Chrulew, and others are turning their attention towards.[3] Chrulew points out that zoos have been bound up with killing on a massive scale, especially with the accumulation of animals "through colonial systems of extraction" resulting in "devastating impacts on habitats and life-worlds"

Figure 11.3 Tiger Enclosure, Boys Looking on. Photo credit: Maurice Frichot.

(141). The history of emergence of the zoo is suspect, and the viewing pleasure it arouses is dubious. It's origin story places it alongside the circus, which then unfolds a narrative of expeditions dedicated to the capture of animals stolen from far off lands, and the dynamic of hunter and hunted. In *Zoo Futures*, Jon Cohen draws attention to the proximity of the zoo to the circus. He notes how, historically, the "line that divides circus from zoo has been murky." Education and entertainment, spectacle and conservation are affectively mixed. The zoo is a theme park as much as it is an institution for edification and conservation. How does memorial culture enter this dynamic? What publics does it produce?

The zoo has been critiqued for its Disneyfication effect, and the spectatorship it invites has even been compared to the consumption of pornography (Acampora). On the other side of the bars, as Chrulew explains, the "befuddlement" of animals, their anxious and "unnatural" behaviours, refusals to breed, and high mortality rate have long plagued the zoo (138). Where developments in zoo design have moved away from the brutality of the caged cell and pit towards live dioramas that ethologically emulate the original habitat of its inmates, violence persists, both hidden and in plain sight, in the biopolitical organisation and spatialisation of the zoo from the cage to the representative patch of territory (Churlew 144). All the while, the scenography of the zoo creates its own memorialising effects, manufacturing the illusion of an animal at home in its habitat, but a habitat secured from prey, from hunger, from broader relational networks of kin and happenstance encounters. We, the spectator, are to be reassured that nothing is out of place, everything is, in fact, in its place on either side of the grille, window, fence. Here, memorial culture risks closing down multi-species relations rather than opening them up. Again, as Parr has argued, memorial culture can settle on fascistic definitions of national identities that nullify difference, as well as lend itself to the spectacle of memory repackaged as Capitalist consumption.

In his famous essay "Why Look at Animals?" John Berger insists that when it comes to the zoo, it is too easy to figure it as a site of marginalisation, such as a prison, a ghetto, a "madhouse," he even cites the concentration camp in this mix (26). The persistent and uneasy question of the ethics of holding other species captive unsettles us when we look at animals who look back at us (Keulartz "Toward"; Keulartz "Captivity"). "The eyes of an animal when they consider a man [sic] are attentive and wary" writes Berger, adding that each regards the other across an abyss of noncomprehension. On either side of a divide asymmetrical power relations pertain. While Berger holds onto the exceptionalism of the human, for his imagined character, "man" becomes self-aware in returning the gaze of the animal and uses language to express this awareness, his sympathies remain with the animals (5). The zoo is a monument to the impossibility of encounters between animals and humans, Berger remarks, instead keeping them apart (21). The denatured man looks back at the bare life of the denatured animal and sees nothing. When Jacques Derrida famously, and hilariously, ruminates on what his cat sees when his cat gazes upon his naked body, the self-awareness of the philosopher is expressed in the recognition of the animal that he, himself, is (372). And yet, his animal self, his self in proximity with the animal, is affected with a sense of shame and embarrassment in his naked exposure. Leading me to ask: What did Lori see when Lori looked back at me?

Reflecting again on my own rescue bird, who was also, effectively, a captive bird, a bird that I was convinced could see me, and was somehow attracted to me (at least as source of food, and as a familiar and comforting being), I remain unsettled by my cross-species encounter. Once Lori had graduated from their breeding box, I set up the rear room of the house, the smallest bedroom, a spare room, as an aviary. It was a short-term plan. The old townhouse was too large and rambling to allow Lori to roam about. And there was the danger of Dog to consider. This solution was better than a cage, but still not suitable. It did not even begin to amply replicate a suitable habitat for a Rainbow Lorikeet. I had to keep reminding myself that Lori was here out of place, divorced from its habitat and comrades. I had constructed what was something of a zoo, a place of enclosure, a preserve, for the time being. All the while, as a family, we were likewise experiencing our own enforced enclosure as lockdown conditions in *naarm* waxed and waned. My attempt at putting my love into practice may well have been misdirected and badly botched. This is something Chrulew warns about (138). Should Lori have been left to its death beneath the Canary Island Palm Tree instead?

While the zoo increasingly takes on the task of caretaker and research partner, supporting efforts to stave off the worst of anthropogenic biodiversity loss and ongoing extinction events, can its past be so easily expunged? "Everywhere animals disappear" Berger writes, arguing that in the enclosure of the zoo, animals "constitute the living monument to their own disappearance" (Berger 26; also cited in Chrulew 137). The zoo's institutional shift towards conservation and education (where education can be extracted from the mere consumption of spectacle) is routinely critiqued, and as it turns out, zoos are simply not large enough nor well-funded enough to construct and maintain a "Noah's Ark" in response to the sixth great extinction event (Rimmel). Berger remarks that the biblical story of Noah's Ark "was the first ordered assembly of animals and man" (19). For whom is Noah's Ark being constructed, for what version of future human or alien other? What coming community awaits the monument that the Ark itself forms in anticipation? Can the zoo be reimagined in its institutional form, or should it be dispensed with altogether as an inherently violent space of enclosure? In the aftermath of our best laid efforts, we will probably not witness the ark as it rediscovers land, disembarking its naïve passengers onto the shores of a new world they have no idea how to negotiate.

Berger remarks that the "the manufacture of realistic animal toys coincides, more or less, with the establishment of public zoos" (23). As our visit to the zoo concludes, we exit via the gift store and purchase a stuffed toy parrot as a memento.

Planetary Monument: Conclusion

Though there are significant exceptions to the rule, the memorial is constructed to impress the onlooker with its scale, gravitas, and grandeur. The memorial is inherently conservative. It aims to conserve the historical record – usually of the victorious – and safe keep memories, aspiring to render concrete intangible events and encounters. As time passes, the memorial will inevitably begin to speak of events that have not been witnessed by those presently gathered before its edifices. Lest we forget. In time, a memorial will contribute to composing new memories for those who come to visit the memorial site. Connection, conjunction, disjunction. Voyages of discovery, colonial misadventures in territorialisation engaging in the expansion of Empire, return with trophies to deposit in grand expositions, in zoological gardens, and in the form of memorabilia that clutter the domestic scenes of the late eighteenth and nineteenth century.

In her work on memorial culture, Parr stresses the role that memorials have on shaping subjectivities, a collective body politic, and even forming a coming community (5; 166–170). A research community, perhaps? The outcome of memorial culture can veer off in radically different directions, manifesting as ameliorative social relations and unforeseeable possibilities or toxic and despotic closures. A line of flight, as Deleuze and Guattari have observed, can result in liberation and joy, or it can result in a line of death (*Thousand* 230). Parr describes the dynamic of desire that animates memorial culture as one where

> memory either affirms, legitimates and advances a paranoid (fascistic) investment of social desire, or it stirs forth an open polyvocal (schizoid) one. The latter is a critical and joyful movement; the former immunizes us against the past while rendering the future mute.
>
> (12)

From the vast point of view of the geological age called the Anthropocene, the whole world might be conceptualised as a vast memorial. Planet earth's lithic crust has been so territorialised, terraformed, and materially redistributed in fervent cycles of dig and dump (Frichot *Dirty*), that we could, from a certain vantage point, argue that we now inhabit a whole world monument. In her discussion of "extinct theory," Claire Colebrook argues that our furious industrial activity has despoiled the world, scarring the earth's surface (10; 23). We have collaborated in composing a planetary graffiti. Humankind has left behind its mark writ large. This returns me to the question: towards whom are our conservation efforts directed, who will receive the Noah's Ark at the end of (human) days?

I return, in closing, to Deleuze and Guattari's counter-intuitive definition of the monument as that which has the potential to compose coming communities. Parr likewise stresses the synthetic rather than retrospective and reflective qualities of what a monument can do. Less remembrance of things past, than memory in the making, or what Parr calls "memory thinking" is what qualifies memorial culture when affirmatively expressed (4). If we take on board this definition of memorial culture as something constructive, then we can see how it might express itself through programs that get the "public" on board in the fight against extinction. There is the sense of a common cause, of a righteous response to the plight of defenceless creatures. Hence the argument that is often made for the continued beneficence of the colonial institution of the zoo. At the same time, we must not forget for a moment that, if these critters are defenceless, then they are defenceless in the face of the ravages that we as a species-being have wrought on a planetary scale. The entanglement of affects, of fighting for a good cause, while recognising our collective culpability makes all these considerations complex. It's not a bad thing to acknowledge this complex, intractable entanglement. Ethnographers such as Tsing and her collaborators find joy here in staying with the mess of contaminated environmental relations. It's about making do, coping, paying attention as best we can (Figure 11.4).

Figure 11.4 Canary Island Palm, Canning Street, Carlton North, Melbourne. Author photo.

In the evenings on the median strip, the Canary Island Palms begin to behave like a chemical clock. First the raucous laughter of lorikeets dominates, then the eerie whining call of the Myna birds in congress. Thinking in flock. It's not the one but the multitude. This collective evening chorus is an expression of species-being. Some months after Lori's departure, still acutely feeling what I can only call a cross-species eco-grief, I'm convinced that the invasive Mynas have taken over by pushing out the local Lorikeets and de-nesting their chicks.

My own feathers get flustered at the thought that I was doing Lori no favours. I continue to achingly wrack my imagination for ways to answer the unanswerable question of what happened to Lori once Lori departed. I'd worked so hard to foster life over the fourteen weeks we shared. The careful preparation of food, the regular feeds, the gradual weaning, the wonder when Lori first demonstrated they could fly, the de-licing regime, the constant measures taken to ensure good hygiene. Then, one late afternoon, enjoying some sun on the upstairs terrace, as recommended for young birds, Lori took flight. Shooting forwards through the air, Lori hooked a sweeping curve around the Canary Island Palm across the way, and then disappeared out of sight. Moments later a flock of Rainbow Lorikeets passed by, and so the story I came to tell myself was that Lori had re-joined their mob and was off and away. Captive born or raised creatures are challenged when introduced to the wild, having no experience in its negotiation, and yet, I held onto hopes that Lori might be the exception. After all, Lori had only sojourned with us a small while. Following all these efforts, Lori and their brood probably cares neither one way nor the other. I will remain haunted by the question, did I succeed in releasing Lori to an unfamiliar wild, or was Lori released to a familiar environment-world? My own desire to memorialise Lori the Rainbow Lorikeet led to the composition of a novella, as yet unfinished, perhaps forever incomplete. Deleuze and Guattari reassure me, though, that a memorial or a monument "may be contained in a few marks or a few lines, like a poem by Emily Dickinson" (*What* 165). Again, whatever material or arrangement is sufficient to hold together that thing we call memory, or, not memory – "Memory, I hate you!" – but an event that has come to pass that has irrevocably transformed all those brought together in the midst of an encounter. The memorials I build to Lori in my imagination, and across the pages of an unfinished novella, are merely the stories I tell myself to soothe myself to sleep amid this sixth great extinction event. They form my own personal memorial. Though it probably matters little, Lori changed me, for I once loved a lorikeet.

Notes

1 I sincerely thank Dejan Stojanovic for speaking to me about his research and work on the conservation of the Orange Bellied Parrot. For more about his research, see https://researchers.anu.edu.au/researchers/stojanovic-d

2 See https://www.difficultbirds.com/study-species#orangebellied-parrot
3 See https://readingzoos.sites.uu.nl/workshops/workshop-4/

Works Cited

Acampora, Ralph R. "Extinction by Exhibition: Looking at and in the Zoo." *Research in Human Ecology*, Vol. 4, No. 1, 1998, pp. 1–4.

Amato, Sarah. *Beastly Possessions: Animals in Victorian Consumer Culture*. University of Toronto Press, Toronto, 2015.

Auden, Wystan Hugh. "Twelve Songs", *The Norton Anthology of Poetry*, 4th edn., Margeret Ferguson, Mary Jo Salter and Jon Stallworthy (eds). W.W. Norton and Company, 1996.

Berger, John. *About Looking*. Pantheon, New York, 1980, pp. 1–26.

Carson, Rachel. *Silent Spring*. Houghton Mifflin Company, Boston, MA, 2002.

Chrulew, Matthew. "Managing Love and Death at the Zoo: The Biopolitics of Endangered Species Preservation." *Australian Humanities Review*, Vol. 50, 2011, pp. 137–158.

Cohen, Jon. "Zoo Futures." *Anthropocene*, 8 March 2013. Accessed 22 October 2022. https://www.anthropocenemagazine.org/conservation/2013/03/zoo-futures/

Colebrook, Claire. *Death of the PostHuman: Essays on Extinction*, Vol. 1. Open Universities Press, Ann Arbor, MI, 2014.

Deleuze, Gilles and Félix Guattari, *A Thousand Plateaus*. University of Minnesota Press, Minneapolis, 1987.

Deleuze, Gilles and Félix Guattari. *What is Philosophy?* Columbia University Press, New York, 1994.

Derrida, Jacques. "The Animal that therefore I am (more to follow)." *Critical Inquiry*, Vol. 28, Winter 2002, pp. 369–418.

Difficult Bird Research Group. Accessed 26 October 2022. https://www.difficultbirds.com/orangebellied-parrot

Dooley, Sean. "The Orange Bellied Parrot." *The Monthly*, April 2008. Accessed 27 October 2022. https://www.themonthly.com.au/issue/2008/april/1273722436/sean-dooley/orange-bellied-parrot#mtr

Engstrom, Sandra. "Recognising the Role Eco-grief Plays in Responding to Environmental Degradation." *Journal of Transdisciplinary Peace Praxis*, Vol. 1, No. 1, 2019, pp. 168–186.

Frichot, Hélène. "Conexa, Conjuncta, Disjuncta: What Can a Monument Do?" *Future Anterior*, Vol. VIII, No. 2, Winter 2011, pp. 74–86.

Frichot, Hélène. *Dirty Theory: Troubling Architecture*. Baunach AADR, 2019.

Hron, Madelaine. "Stillbirth Grief, Eco-grief and Corona Grief: Reflections on Denialism." *Eco-Anxiety and Planetary Hope*, Vakoch, D. A. and Mickey, S. (eds). Springer, Cham, 2022, pp. 67–75.

Keulartz, Jozef. "Captivity for Conservation? Zoo at a Crossroads." *Journal of Agricultural and Environmental Ethics*, April 2015, Vol 28, No. 2, pp. 335–351.

Keulartz, Jozef. "Toward an Animal Ethics for the Anthropocene." *Animal Ethics in the Age of Humans: Blurring Boundaries in Human-Animal Relations*, Bernice Bovenkerk and Jozef Keulartz (eds). Springer, Cham, 2016, pp. 243–264.

Lees, Caroline. and Jonathan Wilcken. "Sustaining the Ark: The Challenges Faced by Zoos in Sustaining Viable Populations." *International Zoo Yearbook*, Vol. 43, 2009, pp. 6–18.

Malabou, Catherine. *Ontology of the Accident: An Essay on Destructive Plasticity.* Polity, 2012.
Miller, Nick. "Proud or Mad? Melbourne's Marathon Lockdown Becomes the World's Longest." *The Age*, 3 October 2021. Accessed 27 October 2022. https://www.theage.com.au/national/victoria/proud-or-mad-melbourne-s-marathon-lockdown-becomes-the-world-s-longest-20210930-p58w9w.html
Parr, Adrian. *Deleuze and Memorial Culture: Desire, Singular Memory and the Politics of Trauma.* Edinburgh University Press, Edinburgh, 2008.
Pihkalu, Panu. "Climate Grief: How We Mourn a Changing Planet." *BBC: Climate Emotions*, 2020. Accessed 27 October 2022. https://www.bbc.com/future/article/20200402-climate-grief-mourning-loss-due-to-climate-change
Rimmel, Gunnar. "Extinction Accounting in European Zoos." *Around the World in 80 Species*, Jill Atkins and Barry Atkins (eds). Routledge, London, 2018, pp. 221–235.
Stengers, Isabelle. *In Catastrophic Times: The Coming Barbarism.* Open Humanities Press and Meson Press, London, 2015.
"The Secret Lives of Our Urban Birds." *Catalyst*, ABC, Accessed 1 February 2022, https://www.abc.net.au/catalyst/the-secret-lives-of-our-urban-birds/13734884
Tsing, Anna. *The Mushroom at the End of the World: On the Possibility of Life in Capitalist Ruins.* Princeton University Press, Princeton, NJ, 2015.
Tsing, Anna, Heather Swanson, Elaine Gan, Nils Bubandt, eds. *Arts of Living on a Damaged Planet: Ghosts and Monsters.* University of Minnesota Press, Minneapolis, 2017.

Index

Note: Page numbers followed by "n" denote endnotes.

For Product Safety Concerns and Information please contact our EU
representative GPSR@taylorandfrancis.com
Taylor & Francis Verlag GmbH, Kaufingerstraße 24, 80331 München, Germany

www.ingramcontent.com/pod-product-compliance
Ingram Content Group UK Ltd.
Pitfield, Milton Keynes, MK11 3LW, UK
UKHW020141200926
475702UK00005B/848

* 9 7 8 1 0 3 2 3 2 6 3 7 5 *